This book is dedicated to Greg Poole,
whose art continues to support conservation
and bring joy to so many people.

First published in the United Kingdom in 2022 by the British Trust for Ornithology, The Nunnery, Thetford, Norfolk, IP24 2PU. www.bto.org

This book was produced by BTO Services Ltd on behalf of the British Trust for Ornithology, Registered Charity Number 216652 (England & Wales), SC039193 (Scotland).

Into the Red © 2022 British Trust for Ornithology, Thetford

Text and Artwork © The Contributors. Dulin photograph by Peter Stone Photography, Swift photograph by Jody Lawrence.

Design, layout, editorial: Mike Toms
Proofreading: Viola Ross-Smith
Lapwings: Will Rose

All rights reserved. No part of this publication may be reproduced, stored in an electronic retrieval system or transmitted in any form or by any means, electronic, mechanical, photocopying, recording or otherwise, without the written permission of the publishers and copyright holders.

Neither the publishers nor the authors can accept liability for the use of any of the materials or methods recommended in this book or for any consequences arising out of their use, nor can they be held responsible for any errors or omissions that may be found in the text or which may occur at a future date as a result of changes in rules, laws or equipment. Opinions expressed by *Into the Red* contributors do not necessarily represent the views of BTO or the RBBP's funding partners, and the Rare Breeding Birds Panel does not endorse these opinions.

ISBN 978-1-912642-38-0

Printed and bound in Italy by Printer Trento, an FSC® certified company for printing books on FSC mixed papers in compliance with the chain of custody and on-product labelling standards. Printer Trento has an ISO 14001 certified environmental management system.

INTO THE
RED

Curated by Kit Jewitt & Mike Toms

Foreword
Kit Jewitt

Casting my mind back to Valentine's Day 2020, the publication day of our last book, *Red Sixty Seven*, I would be surprised if many of us could have predicted what was to come.

Within a few weeks our lives would change dramatically with everyone being ordered to stay at home, distanced from our loved ones to ensure we saved lives. But whilst we were locked down in our homes, isolated from friends and family, the pandemic somehow brought many of us closer to nature. Recalibrating our senses around our shrunken, quieter lives allowed us to hear voices that had been drowned out by the white noise of daily life. And we listened, and we heard them. The birds. And their voices brought light in the darkness to so many of us. It was a welcome reminder that they were there but, for some of them, a warning that they might not be for much longer. For some of these birds their numbers are declining so rapidly, and their world is so diminished, that they are living on a wing and a prayer.

Red Sixty Seven was born out of necessity and a desire to do something. A book of art and words to raise awareness and funds to help address the declines of the sixty-seven species at that time on the UK Red List of Birds of Conservation Concern. Thanks to the wonderful artists and writers in the book, and those who bought a copy, it was a great success, with profits amounting to over £40,000. And every penny has been channelled into projects to help the bird species freefalling fastest, such as our breeding waders, seabirds and the iconic Cuckoo.

Last year, thanks to new research evidence and data, the list was updated in its fifth iteration. It now holds 70 species of greatest concern. Joining birds such as the Starling and Puffin on this most ignominious of lists are several new, yet very familiar additions, such as the Greenfinch and House Martin, and the bird I wait for each year with feverish anticipation more than any other; the Swift. I tilt my ear to the heavens, to help me hear their heart-piercing screams, squinting the sun from my eyes to scan for the Tie Fighter-like squadrons of these otherworldly creatures tearing through the skies. But each year their numbers seem fewer, and their visits seem briefer. It's July now, and I can count on one hand the screaming Swifts I've heard over my garden, excluding the ones played from my phone on a bedroom window ledge, desperately advertising the two perfectly formed starter homes I attached to the back of our house for them last year. These boxes may not have attracted any passing Swifts, but at least my local House Sparrows, another Red-listed species, have benefitted, turning them into luxury penthouse apartments.

The houses opposite me used to boast a thriving colony of House Martins when we moved here over a decade ago, but the skies have been absent from their vocal flatulence for a few years. And I only occasionally hear the wheezy call of a Greenfinch now, when I recall they visited my feeders regularly when we first moved in. Hedgerows where I used to hear Yellowhammers shouting about 'bread and cheese' are silent. I'm sure many of you reading this will have similar stories and experiences of the insidious but perceptible changes happening to the once common birds on our doorsteps.

And then there are the extinctions. Happening under our noses, in our lifetimes, right now. The updated Red List shows that one species, from surely the most pored-over page of many a child's book of British birds, the enigmatic Golden Oriole, is now extinct as a UK breeding bird and joins the Wryneck, Temminck's Stint and Serin as having dropped off the list altogether. And it's highly unlikely they will ever return. At least we can save money on overpriced petrol on annual pilgrimages to RSPB Lakenheath in Suffolk.

But isn't all bad news, as the White-Tailed Eagle has proven. Successful reintroduction schemes have allowed this species to move from the Red to the Amber List. But the successes are few. There is still much to do.

So, casting aside any fears of the 'difficult second album', we dusted off our address books, polished our brass necks and got back to work. And thanks once again to the kindness of fellow bird-lovers, we have curated the book you now hold in your hands. And once again all of the profits raised from sales of this book will be used to help the birds in its pages, and prevent others from slipping 'Into The Red'.

And hopefully, together, we can give the birds that we love a fighting chance.

Beyond just a wing and a prayer.

Introducing the Red List
Mike Toms

The Red List contains those bird species that have been identified as being the most vulnerable; these are the ones whose populations have shown the greatest levels of decline or whose numbers are so small that they warrant our attention. But how do we determine which species should be placed on the Red List, and why does listing matter?

Older birdwatchers often remark on the changes that they have seen over the years; species once common that are now only rarely encountered, or others that were once rare visitors but which have now become familiar residents. While such remarks typically reflect genuine patterns of change in bird populations, they are also shaped by our own shifting experiences. Perhaps the birdwatcher grew up in one part of the country but now lives in another; perhaps they now visit different places, or visit the same places but at different times. If we are to identify real change then we need hard evidence, the sort of evidence that comes from data collected in robust and repeatable ways.

The data that we have on the UK's bird populations tell us a lot about how these populations have changed over time; for most species this means since the 1960s but for a few, such as Grey Heron, our data go back further. We have information on changes in both abundance and distribution, together with additional data on some of the things that drive these changes, such as breeding success, survival rates and movements.

Collectively, these data can be used to identify which species are most in need of conservation action, something that is done here in the UK through periodic assessments that are carried out by a panel of experts using standard criteria. This is known as the Birds of Conservation Concern process. Identifying the species that are most in need of our help ensures that the limited financial resources available for biodiversity conservation can be directed to where they are most needed. It is essential that such assessments are based on the best available information, and that the assessment process itself is transparent and robust. It is also essential that the results of the listing process then feed through into policy and action.

The Birds of Conservation Concern assessments use a suite of quantitative criteria and apply these to all of the UK's regularly occurring species. The approach uses a simple traffic light system – Red, Amber, Green – as a means of classifying the conservation status of each of the species assessed. There are 'Red' criteria with thresholds for rates of decline in numbers and range, historical decline and international status. Then there are

'Amber' criteria for the same trend measures, plus some other considerations, including rarity and how localised a species is. Each species is first assessed against the set of Red List criteria; if it meets any of these then it is Red-listed; if not, then it is then assessed against the Amber List criteria. If it meets these, then it is placed on the Amber List. If it does not meet any Red or Amber List criteria then it is placed on the Green List.

The periodic assessments draw heavily on the wealth of data collected by the many thousands of volunteers who participate in the national monitoring schemes operated by the British Trust for Ornithology (BTO), Royal Society for the Protection of Birds (RSPB), the Joint Nature Conservation Committee (JNCC) and others. The first Birds of Conservation Concern review was published in 1996, the latest in December 2021.

Currently, there are 70 species on the Red List (nearly double the figure from that first review in 1996), 103 on the Amber List and 72 on the Green List. Since the last review, carried out in 2015 and on which *Red Sixty Seven* was based, the Red List has grown by three species and the Amber list by 11. One species, the Golden Oriole, has dropped off the listing altogether, because it is now considered extinct as a breeding bird here in the UK.

The growing length of the Red and Amber Lists is a cause for concern, as is the underlying pattern of species present on these two lists. Farmland and upland breeding birds are particularly evident in these lists, as are those summer migrants that winter in Africa. Worryingly, concerns over the status of our wintering wildfowl and wader populations have also increased following the latest assessment. However, it is not all doom and gloom. Thanks to targeted conservation action, we have seen White-tailed Eagle move from Red to Amber, following in the footsteps of Bittern, Stone-curlew, Woodlark and Red Kite, and something that demonstrates that, with the required resources, we can turn things around.

The Rare Breeding Birds Panel

Dr Mark Eaton, RBBP Secretary

The Rare Breeding Birds Panel (RBBP) was established as an autonomous body in 1973, in order to "collect in one place all information on rare breeding birds in the UK so that changes in status – both increases and decreases – can be monitored". Nearly 50 years on, the Panel maintains that purpose, although over the intervening decades it has, of course, become more sophisticated in how these data are collated, archived and reported.

The main functions of the Panel are to maintain the definitive archive of all rare breeding birds in the UK, to publish an annual report on numbers, distribution and trends, and to make available both source data and derived information to support conservation action. RBBP is independent, although it includes representatives from RSPB, BTO and the Joint Nature Conservation Committee (JNCC; on behalf of the Statutory Nature Conservation Bodies), as well as independent members chosen for their expertise in the monitoring of rare breeding birds in the UK. The bulk of the Panel's work is done by a professional Secretary, whose post is co-funded by the RSPB and JNCC, with additional funding to support the running of the Panel provided by BTO. The other members of the Panel advise and support the Secretary.

The Panel works by receiving and collating all available data on breeding bird species with fewer than approximately 2,000 breeding pairs in the UK (also including some races, such as the Fair Isle Wren and Blue-headed Wagtail). The list covered thus includes all the UK's regular rare breeding species, from extreme rarities such as the Bluethroat up to more widespread but still scarce birds such as Avocet and Little Egret. It also includes species which have bred occasionally or have shown breeding behaviour although are yet to breed; some of these species might be future colonists. In addition, the Panel also collects data on rare non-native breeding species (with fewer than 300 breeding pairs).

The work is technically complex because, unlike many other national monitoring schemes, there is no simple, single or structured source of data. The principal source of data is via the network of 80 county and regional recorders. Birdwatchers and other members of the public are encouraged to submit their records to these recorders, either directly or via online reporting portals such as BirdTrack (birdtrack.net), to enable them to compile the relevant data to submit to the RBBP Secretary each year. The Panel also receives data directly from many other sources including RSPB reserves monitoring, species recovery projects, the BTO/JNCC Nest Record Scheme, BTO/

JNCC Ringing Scheme, returns for activities undertaken under Schedule 1 licences, raptor and seabird monitoring, and single species studies and surveys. The great majority of these data arise from the efforts of volunteer ornithologists – the RBBP is an excellent, long-running example of citizen science, dependent on the skills, experience and enthusiasm of thousands of contributors. The resulting data provide critical information, not only about changing status of the birds themselves, but also about their key sites; the RBBP data thus contribute to the management of these sites.

The Panel works to improve the recording and reporting of rare breeding birds by providing encouragement, guidance, and expertise. In recent years we have been partners in national surveys of Red-listed species such as Turtle Dove, Whimbrel and Willow Tit. We also work to encourage responsible behaviour in relation to visiting and sharing information on the breeding locations of rare breeding birds.

The data the Panel collate are maintained in a secure archive accessible only to the Panel's Secretary, and are made available, with appropriate checks and restrictions on dissemination, for a range of research and conservation uses. They are also used by government as a component of Official Statistics relating to state of environment reporting; in national reporting to multilateral environmental agreements, like the Convention on Migratory Species, as part of UK obligations under such international treaties and to help inform implementation; and for assessment of the status of protected areas. A comprehensive annual report, published in the journal *British Birds*, summarises the annual status and trends of all rare breeding birds in the UK for the given year. Past reports, as well as a wealth of other information, can be found on the RBBP website at www.rbbp.org.uk.

Of course, the Panel's data play an important role in Birds of Conservation Concern assessments, enabling the status of rare breeding birds to be assessed robustly. RBBP data holdings provided population trends for over 50 species in the latest Birds of Conservation Concern assessment. Unsurprisingly, many rare breeding bird species are of conservation concern; 23 of the species that the RBBP reports on regularly were Red-listed in Birds of Conservation Concern Five. Two of the species that moved to the new Red List, Purple Sandpiper and Montagu's Harrier, are amongst the UK's very rarest breeding birds. Other Red-listed species, such as Turtle Dove and Willow Tit, are more abundant but in rapid decline.

British Trust for Ornithology

Mike Toms, BTO Head of Communications

The British Trust for Ornithology (BTO) sits at the heart of bird monitoring and research, its partnership of professional staff and many thousands of volunteers delivering the robust science and evidence base that underpins policy decisions and conservation action, both here in the UK and more widely.

Since its formation in 1933, the organisation and its volunteers have tracked the changing fortunes of our bird populations, alerting us to the impacts of habitat loss and change, shifting agricultural practices and a changing climate, all of which have impacted our breeding and wintering bird populations. BTO data have, for example, revealed the consequences of agricultural intensification on farmland birds, the shift in the timing of breeding driven by climate change, and the impacts of emerging diseases, like trichomonosis. These data have also seen species placed on the Birds of Conservation Concern Red List, and shape the opportunities we have to identify and test solutions, through which we can secure a better future for the UK's birds.

BTO's long-term data and rigorous scientific analyses have never been more important, their use today helping us to address the urgent global crises of biodiversity loss and climate change. These data, coupled with the charity's distinctive approach, provide the impartial, impactful and evidence-based knowledge that empowers society to respond effectively to these and other challenges. BTO work on farmland birds, for example, hasn't just alerted us to the problems faced by species like Skylark, Tree Sparrow and Corn Bunting, it has also tested conservation solutions and informed policy tools designed to reverse the declines seen. BTO's ongoing monitoring is being used to determine whether the solutions identified, and the policy tools, used are delivering their intended conservation benefits.

What makes BTO different is the way in which the organisation builds and harnesses the passion, knowledge and curiosity of thousands of people to tackle the urgent challenges of today, advancing our shared understanding of birds and their habitats. Each year, BTO volunteers contribute in excess of two million hours to the collection of ornithological data, representing an in-kind contribution of £41 million annually. This generosity delivers a significant part of the UK's bird monitoring, and underlines the central place that birds hold in the affections of so many people, from many different walks of life. Contributing to BTO's work is a positive step that individuals can take for themselves, helping to build a better future for birds, wildlife and people.

As noted by Mark Eaton in the previous section, the monitoring schemes operated by BTO, typically working in partnership with other NGOs, also feed vital information through to where it is most needed. BTO's extensive work on seabirds and their interactions with marine renewables is being used to inform decisions about where renewable energy schemes are placed, and how they are designed. Similary, data on the distribution and status of bird populations around our coasts and inland are routinely used to designate sites as being of particular conservation importance.

There is another aspect to BTO's work that deserves a mention here, in the context of this book, and that is the organisation's role in inspiring people to care about birds and what is happening to them. The scientific outputs that feature in peer-reviewed papers and reports provide the hard – often stark – evidence of what is happening to our bird populations. But just producing the statistics is not enough. An increasing body of research suggests that scientific messages may be communicated more effectively through the use of art, storytelling and photography, rather than via peer-reviewed papers. The need to broaden engagement with work on declining populations of migrant birds first prompted BTO to explore the potential for an arts and culture approach to better present its ongoing research into UK birds that winter in Africa. This was achieved through its 'Flight Lines' project, which saw artists working alongside BTO researchers and volunteers to tell the story of migrant birds and the work being done to address the decline evident in their populations. Audiences exposed to scientific ideas through such pathways have been shown to demonstrate meaningful change in their understanding of topics and, where the topic has an environmental basis, their behaviour.

The success of this approach has been followed up through other projects, including *Red Sixty Seven* and *Into the Red*, again using the creative arts to engage new audiences with our work. Birds inspire us, and they are accessible to just about all of us – you only have to look up to see a bird! If we are to secure a better future for them, and for our shared environment, then we need to continue to reach more people and inspire them to take action. By participating in BTO's monitoring work, collecting the evidence that is needed to support good decision-making, individuals become part of the solution, working together to deliver a world that is better for birds, wildlife and people.

Acknowledgements

This book would not have been possible without the generosity and creativity of the 140 contributors, each of whom so kindly gave their time, passion and skills to this project. It has been a joy to work with you all, to see before anyone else the results of your creativity, and to share with you the creative energy that has driven this project through to completion. In addition to those whose names appear throughout this book, we wanted to say an extra thank you to Hannah Athayde, Helen Baker, Amy-Jane Beer, Paul Burger, Jo Clark, Mary Colwell, Jamie Dunning, Ben Hoare, Michael Hogben, Gayatri Kaul, Steven Lovatt, Kirsty McLachlan, Gill McLay, Lucy Pedlar, Martina Pizzini, Becca Ptaszynski, Will Rose, Viola Ross-Smith, Ben Rothery, Lisa Thomas, Abi Warner, Mark Whyles and Jenna Woodford.

INTO THE
RED

Grey Partridge

If there was ever a bird that was our Canary in our cornfields, then the Grey Partridge has this unfortunate accolade. Its decline is well documented in game books across the country where bag sizes reduced significantly in less than half a decade. Following the Rio Earth Summit in 1992, the UK government's response to the Convention on Biological Biodiversity was to produce the first National Biodiversity Action Plan. Unsurprisingly, the Grey Partridge was included in this first list.

The rapid decline of this charismatic gamebird can be linked inextricably to the changes in agriculture. Almost overnight, spring cereals were replaced with winter cereals and the summer scenes depicted by Thorburn, of Grey Partridges foraging with their young in green post-harvest stubbles, soon became assigned to paintings and picture books.

I am lucky enough to have encountered these arable specialists on numerous occasions and am always impressed by their parenting skills, devotion and attention. Both members of the pair will alternate incubation and sometimes there may be two nests on the go. The female's plumage will turn much darker during incubation, close to black. The hatching is timed with the longest hours of daylight and once hatched the young balls of fluff, no bigger than a bumblebee, are incredibly mobile. Within 10 days they master short flights and at three weeks are accomplished fliers. This rapid growth is only achieved through the ingestion of quantities of suitable invertebrates; as many as 2,000 in one day per chick, and with the larvae of sawflies particularly nutritious. Both parents defend their young with vigour, the female guiding the brood to better foraging while the male follows close by, always on alert. Standing tall and neck stretched high with eyes peeled for predators, both ground and sky.

All this protection and care will come to nothing if there are insufficient quantities of suitable food. The rapid growth of the young can only be achieved on this high protein diet. When food is scarce the daily forage is further protracted and the young birds soon become weak and prone to cool and inclement weather. Broods can number as many as 20 plus to start with, but soon dwindle without sufficient nourishment. With the average age of an adult being not much more than 18 months, high productivity is key and, as we saw in the 1970s, decline can be rapid.

CHANGING BOCC STATUS 1 2 3 4 5
RED-LIST CRITERIA APPLICABLE
- IUCN Globally Threatened
- Historical breeding population decline
- Breeding population decline
- Non-breeding population decline
- Breeding range decline
- Non-breeding range decline

WORDS – JAKE FIENNES | **ART –** ANDREW HASLEN

Ptarmigan

High above Loch Tay in the southern Highlands stands the ridge of Meall nan Tarmachan – 'the hill of the Ptarmigan'. 'Tàrmachan' translates from Scots Gaelic as 'croaker' or 'murmurer' and beautifully describes the evocative creaking call that you can hear echoing across the corries in spring.

Our only truly montane bird, these decidedly tough wee birds spend the entire year on our highest and most exposed habitats. They are fantastically well adapted to life in the mountains, famously moulting from mottled grey summer plumage to white winter attire each autumn. In many ways this is part of the problem for Ptarmigan in modern-day Scotland. Their moult is triggered by shortening days but all-too-often they are almost entirely white when their mountain home has little to no snow cover. This makes them easy pickings for upland predators.

I've been lucky to have spent many days amongst Scotland's high mountain tops in all seasons. In mid-winter you can find yourself battered by ferocious winds, clad head-to-toe in layers of technical clothing and having to wear ski goggles just to see. It is truly humbling then to stumble across a covey of Ptarmigan nonchalantly feeding and wandering about, seemingly unaffected by the conditions. In contrast, I vividly remember being at the top of the Devil's Point in the Cairngorms one beautiful day in early summer. I found a female Ptarmigan with well-grown chicks in tow by the summit, a world away from the conditions these chicks would face come winter.

Ptarmigan can be notoriously hard to find, not least because of their high-altitude haunts and effective camouflage. I remember one particularly comic moment searching for them on a hot sunny day, guiding a group of birders to look for them in the Northern Corries. Such weather forces these well-insulated birds to seek shade, so we concentrated our efforts scanning a huge boulder field out of the sun. After some time we heard the familiar creaky call and began to scrutinise the grey landscape for a grey bird. "I've got one!" someone called. "Whereabouts?" asked the rest of the group "Ummm, they're up from that huge block, by that grey boulder, the one that's shaped like a … wait a minute that's another Ptarmigan!" and so it went on; when we finally got them in the scopes we were treated to rewarding views, well earned.

These days we have lost Ptarmigan from the Southern Uplands, Orkney, and all but a few of the Western Isles. There are still Ptarmigan to be found on Meall nan Tarmachan today, but it would be tragic if their last legacy were a hill name and the memories of a few lucky hillwalkers.

WORDS – STEVE WILLIS | **ART** – PAUL WARD

Capercaillie

As mossy and ancient as the Caledonian Forest, the Capercaillie is a species that seems too improbable to exist on our 21st century island home. It was once described to me as a "masterclass in hiding a giant turkey in a forest". For years, this was how this bird taunted me: a turkey-sized ghost, hiding, in spite of its enormous size, within a maze of resinous pine and Blaeberry.

Eventually, I would use careful fieldcraft, placing my camouflaged tent early the evening before, in a secluded area, to carefully observe the few, extraordinary moments when the 'Horse of the Forest' reveals itself. I will never forget the first morning that I got it right. A guttural sound of popping corks and scratching filled the dark air as I awoke. Moving within my hide at a pace that a snail would find embarrassing, I rose to the window, and there, in the gloaming, were the giant shapes of male Capercaillies, each walking around, tail raised, in their own, self-important world of pomp. Capercaillie leks are unreal; to look out, and think that this bizarre scene was still playing out in modern Scotland, was unreal too. It was like a window back into the early Holocene. Nothing in this drama had changed since then.

These leks, hidden within forest clearings or on quiet forest tracks, are nature's most easily disturbed coffee mornings. Capercaillies are adapted to huge tracts of quiet, undisturbed woodland. They have clung on at the edge in Britain for a long time, but as more and more people walk, unawares, into their leks, or cycle, or walk their dogs through them, finding the quietude to lek in peace has become an ever larger problem; disrupting the vital social structures on which successful breeding depends.

Whilst wet summers and rising Pine Marten predation have been blamed for Scotland's Capercaillie decline, the same factors are present in Scandinavia, where Capercaillies remain so common in parts of Sweden and Norway that hens with their broods can be seen along roadsides as we might see Pheasants here in the UK. Can we blame weather and the predators alongside which these birds evolved? Another perspective would be that we simply do not have enough Caledonian Forest, at the moment, for a bird that needs space to survive. In the Cairngorms, conservationists and landowners are working on changing all of this; regrowing an ever larger forest, a more natural forest, for the future, and using cattle once more, to promote a better ground layer for nesting hens and their chicks. With concerted efforts, it must be hoped that the Capercaillie can survive the coming decades – to pop like a champagne cork at a time when most of us are tucked up in bed; dreaming, perhaps, of finally discovering a giant turkey in a forest.

WORDS – BENEDICT MACDONALD | **ART** – FEDERICO GEMMA

Black Grouse

The bubbling call was as evocative in real life as the recordings I'd heard. I felt tears in my eyes as I cautiously peered through the netting of the hide and caught a glimpse of the males. I didn't need binoculars, or a scope. I felt almost part of the lek, with these males charming me as much as the female grouse that they were trying to impress.

Black Grouse are one of the iconic birds of upland Scotland and Wales, although they should be more widespread across much of the UK, especially in England where their historic range reached as far south as Hampshire, Cornwall and Norfolk. Population declines, even in their upland strongholds, are primarily a result of habitat change.

The males, or cocks, are striking – black with red eyebrows, white wing stripes and a white plume of feathers in their tail. The females, or hens, are better camouflaged – a mottled brown colour which makes them much more difficult to spot, even at a lek site, where they will only regularly make an appearance during the mating season. When females are present at the lek the males will frequently fight, viciously at times, to win the chance to mate. Throughout the rest of the year the males come together to size each other up, competing with one another in non-violent display for dominance.

And this was what was happening as I peered through the hide at a Black Grouse lek in Scotland, where I was part of team filming for BBC's Springwatch in 2021. We were operating as a very slimmed down crew to minimise disturbance and were sitting in silence, moving as little as possible. It was late in the season to see so many birds together but the cold spring ensured that they were still as competitive as ever.

I'll never forget the morning, the literally freezing air, the trek to the site in darkness and the silent wait for that magical moment when the lek started. It began with a hissing "*pwssshh, pwssshh*" sound, and was soon followed with the classic bubbling calls I was expecting. Seven males were there in all, posturing and showing off to one another with their tails spread wide and their white tail feathers dazzling. I waited until we'd filmed everything that we needed before taking the opportunity to dare a look myself; that moment stands out to me as one of the best birding experiences I've had.

CHANGING BOCC STATUS 1 2 3 4 5
RED-LIST CRITERIA APPLICABLE
- IUCN Globally Threatened
- Historical breeding population decline
- Breeding population decline
- Non-breeding population decline
- Breeding range decline
- Non-breeding range decline

WORDS – RUTH PEACEY | ART – TONY BLYTHE

here we go...it's that time of year when he starts making weird sounds and showing off...

Bewick's Swan

Many moons ago, I had never heard of the Bewick's Swan. Then, back in 1977, I was invited to interview for a lowly Research Assistant post continuing the long-term study, initiated by Sir Peter Scott, of the flock wintering at Slimbridge.

Fortunately, in a pre-Google era, Myrfyn Owen's *Wildfowl of Europe* book was on sale in the Leicester University bookshop. This revealed that Bewick's Swans are long-distance migrants which spend the summer at high latitudes in the Russian Arctic. Also, perhaps more relevantly, variation in their bill patterns enables trained observers to identify individual birds within a flock, without the need for artificial markings. I was hooked!

During my first winter at Slimbridge, I soon became drawn into the lives of these birds. Being able to recognise individuals at a glance quickly resulted in them becoming old friends, leading to concern if a particular bird went missing for several days. Bewick's Swans are faithful not only to their mates but also to favoured wintering sites. This leads to keen anticipation each year as to which birds will return safely, whether singletons will have found a mate during the summer, and which pairs will be accompanied by cygnets.

For most of the 20th century, little was known in the west about conditions on the swans' breeding range. Their lives were a mystery for half the year. Political change in the Soviet Union during the early 1990s, however, facilitated collaborative expeditions, and these proved a revelation. In contrast to the managed landscapes of Western Europe, during the summer the swans inhabit the vast Arctic tundra. Here lakes, pools and channels provide an abundance of suitable breeding territories; open views enable nesting pairs to spot predators from afar, and the long days maximise time available for feeding. Non-breeders continue to congregate in flocks but defence by pairs of their territories is vigorous, with their aerial battles showing amazing agility in excluding other swans from their patch.

The Bewick's Swan is much-loved internationally, and the declining numbers are of great concern. The joy gained simply from watching them must not be lost, and efforts to conserve them are underway in all countries along their flyway. Their graceful evening flight to a winter roost, gentle "*bewking*" calls maintaining contact during migration, and patience in tending their nest and young, will then provide inspiration for generations to come.

CHANGING BOCC STATUS 1 – 2 – 3 – 4 – **5**
RED-LIST CRITERIA APPLICABLE
- IUCN Globally Threatened
- Historical breeding population decline
- Breeding population decline
- Non-breeding population decline
- Breeding range decline
- Non-breeding range decline

WORDS – EILEEN REES | **ART** – CELIA HART

White-fronted Goose

Walking the mile from base camp to our small wooden hide through the whirling evening snow, the valley was filling with deep drifts. After six snow-free weeks, the gale force blizzard of the previous day was now subsiding. Summer 'nights' in west Greenland – where we were observing incubating White-fronted Geese – were scarcely dark. As the snowfall lessened, I searched the white landscape in vain for the goose nest.

Slowly as the snow stopped, about midnight, the clouds began to thin. An orange glow appeared to emanate from the fjord peaks; a very pale blue and pastel pink sky above them. Sharply contrasted against this were the stark black and white cliffs and slopes. For some hours, I fruitlessly scanned as the sun rose. Suddenly, the incubating female erupted from the deep snow, shook herself vigorously, before resettling on her nest, presenting the lowest possible profile to predatory Arctic Foxes.

On 21 June she hatched six goslings and, with her mate, led them three kilometres up the steep valley sides and onto the lake-studded plateau used for brood-rearing and moulting. In late July we rounded up and marked this and other families, now flightless; geese subsequently seen throughout their global wintering range – from Orkney in the north, south to Wexford Slobs in southern Ireland.

Study results from Greenland, along with those from the Icelandic spring and autumn staging areas, and the wintering sites in Ireland and Britain, enabled the development of effective flyway-wide conservation measures. Yet, despite increasing numbers after protection from shooting came in 1982, numbers have been in sharp decline since 1992; the result of too few young being produced annually to replace dying adults. The cause relates to warming of the North Atlantic and the heavy snow of recent springs; these no longer arrive during incubation but come earlier, as the geese arrive from Iceland and need to feed intensively to lay a clutch of large eggs.

White-fronted Geese of another race reach England's shores from Arctic Russia. Falling national numbers of these birds (explaining their Red List status) is not due to population decline (more than a million winter elsewhere in Europe), but the milder winter conditions in continental Europe, where the geese now stay, much to the chagrin of farmers there. Both populations are declining because of the changing climate, though for different reasons. Effective conservation interventions require international coordination and long-term monitoring. The conservation of migrant birds is never simple!

CHANGING BOCC STATUS 1—2—3—4—5
RED-LIST CRITERIA APPLICABLE
- IUCN Globally Threatened
- Historical breeding population decline
- Breeding population decline
- Non-breeding population decline
- Breeding range decline
- Non-breeding range decline

WORDS – DAVID STROUD | **ART** – NICK HAYES

Long-tailed Duck

Now you see them, now you don't: up for air, and they're back down again, feathers pressed tight against dense bodies before each plunge into brisk northern seas. Feet large, wings compact, like the other diving ducks; yet their constant disappearing act while foraging earns them the record amongst their peers for most time spent under the surface, a one-to-four ratio of breath to brine. What sleight of wing! For theirs are wings that can – most unusually – propel them through water and dare them down, deep and deeper, to an astonishing 55 metres beneath the waves, another superlative for family Anatidae.

Don't let the uninspired moniker fool you. There's something so punk about the Long-tailed Duck, especially the midwinter male, with his bleached mohawk above a high dun-coloured forehead, white epaulettes streaked dramatically over his chocolate black back, and a rebellious slash of pink on his bill. Two flamboyant tail feathers stick up and out, impractically elongated. But thoroughly describing the costume of the Long-tailed Duck is a complicated affair, as males differ from females and each has not two but three seasonal runway looks. Their plumage is not to be pinned down.

In Iceland – where I once lived, and where many of the Long-tailed Ducks wintering in British waters return every spring to breed in sub-Arctic wetlands – it's not the look of these birds but their distinctive voice that gives them their name, 'hávella'. Notorious for their garrulousness, they group in raucous flocks where females' quacks are all but drowned out by the males' unconstrained yodelling. According to tradition, the rowdier the 'hávella', the wilder the weather to come; yet these are perpetually rambunctious ducks, calling out day and night to rhythms of their own.

Now you see them, now you don't: Long-tailed Ducks, amongst the most populous Arctic and Baltic birds, sound every part the erstwhile Passenger Pigeons of the North, their vast rafts blanketing frigid seas. And yet, their once death-defying deep dives are, in these times, deadly, as they risk plunging into fishing nets – lively, convivial birds reduced to bycatch. Unceremoniously tangled in gillnets rather than essentially entangled in marine ecology, these once superabundant birds are on the wane. Their decline spells top-down crumbling of ecosystem processes and consequently, ironically, the disappearance of the very species the fishing industries seek. Have you seen them, in their prodigious flocks? What a privilege of boreal extremes – just don't blink.

CHANGING BOCC STATUS 1–2–3–4–5
RED-LIST CRITERIA APPLICABLE
- IUCN Globally Threatened
- Historical breeding population decline
- Breeding population decline
- Non-breeding population decline
- Breeding range decline
- Non-breeding range decline

WORDS – SHAUNA LAUREL JONES | **ART** – PETER PARTINGTON

Velvet Scoter

Each August, the world's entertainers descend on Edinburgh, one of its most beautiful cities. I have often visited at this time of year to stay with my theatre-loving uncle David, who joins me one morning for the short bus ride east to Musselburgh. This is where the river Esk sends out its rich mud into the Firth of Forth, enticing one of just three flocks of Velvet Scoter that summer in the seas around Scotland.

Elsewhere in these miraculous waters is a small band of visiting Red-necked Grebes and, lost among the crowd of commoner cousins that engulf him, a young King Eider coming into his cubist good looks. Never mind the comedians, drag queens and dancers up the road, down here the seaduck circus has come to town.

And yet. There is no better hiding place than a great stretch of grey water. I live far inland and the scale of the sea relative to the birds that live on it always comes as a surprise. Any Velvet Scoters we do find today will be dots through the drizzle, I realise, not the dazzle of black, white and orange the internet and I have promised. Perhaps I should have warned David, on his first such expedition and raising a sceptical eyebrow, that our prospects of finding one at all were similarly unspectacular.

The rocky Baltic shores of Finland and Sweden are home to most of Europe's Velvet Scoters during the breeding season. Others nest halfway up Norwegian mountains or in clearings among the conifer forests of northern Scandinavia, before wintering out at sea, but numbers are only half what they were just 40 years ago. Those that manage to avoid entanglement in fishing nets or poisoning by industrial pollutants may still go hungry, their mollusc prey gobbled up by hungry trawlers. The few hundred Scottish summer birds, which gather here to moult, are joined later in the year by a British winter population of around 3,000 individuals. In the 1980s, three times as many could be found in the Moray Firth alone.

Two scoops of Irn-Bru ice cream but not a lot of birds later, David has wisely caught the bus back home. I am still on the beach, where three Red-necked Grebes reveal themselves, momentarily, through the telescope of a magnanimous observer I have been lucky to encounter. Could he conjure up some scoters, too? The answer, alas, is no. But they are out there somewhere, just beyond the haar, like warm black jewels of the deep liquid murk.

CHANGING BOCC STATUS ❶-❷-❸-❹-❺
RED-LIST CRITERIA APPLICABLE
● IUCN Globally Threatened
● Historical breeding population decline
● Breeding population decline
● Non-breeding population decline
● Breeding range decline
● Non-breeding range decline

WORDS – TOM STEWART | **ART** – JOSÉ ANTONIO SENCIANES

Common Scoter

It's early May and the rush of meltwater in the Pasvik taiga has become an accompanying white noise for the chatter of hundreds of newly-arrived Bean Geese. They sit stiff-necked and alert, eyeing me from ice floes which straddle the river, demarking Norway's soil and the Russian forest on the opposite shore. Spring thunders into this part of northern Europe as the sun, throwing away its circadian rhythm, bounces higher and higher above the horizon with every passing night, until daylight reigns constant. Winter's deep cocoon of snow quickly recedes; a blank canvas soon scribbled – blooming rivulets revealing fragrant layers of Northern Labrador Tea, prostrate *Salix* and *Sphagnum* mosses; an irresistible orange and purple ombré.

The breeding birds rapidly follow. Among the lumps of ice and deadwood, a circle of Goldeneyes are displaying in the watery no-man's-land. Distantly, a Moose grazes, and then a small flotilla of dark ducks edges around a bend in the current – I lift my binoculars and gasp. Common Scoters!

Some years ago, a couple of friends and I embarked on curating the 'Everything List' – a jab back at the manner in which our fellow birders devalued species once 'ticked'. We fought against a culture of no longer 'needing' a bird and the jokes of pages torn from *Collins Bird Guide* once 'done'; an ornithological "Completed it, mate". How can you begin to pretend that you know something intimately if you've not seen all, nay most, of its iterations of plumage, its stages of migration, of breeding, feeding, posturing and posing? We abandoned the idea once it got to documenting moult stages of gulls, but the theory remains part of our birding psyche. And up until that moment on an icy river at 69° north, Common Scoters were one-dimensional background birds for me; to sift through for white wing patches or the low-slung shape of a loon. Tallies on a sea-watch between the highlights. My Everything List entry for the species was meagre. But here they were, in the land of the Siberian Jay and the Northern Hawk Owl – Common Scoters! In the Arctic! In a forest! It was like being reacquainted with old friends, but simultaneously meeting them for the first time.

We're lucky to still have a few pairs of Common Scoters breeding in Britain, clinging on in Scotland's flow country, where they raise their dusky ducklings each summer. The species used to nest across greater swathes of northern and western Scotland, but has faced significant modern declines. Never dismiss those lines of blank, black ducks on your sea-watch again. Think where they hatched, think where they're heading!

CHANGING BOCC STATUS 1–**2**–3–4–5

RED-LIST CRITERIA APPLICABLE
○ IUCN Globally Threatened
● Historical breeding population decline
● Breeding population decline
○ Non-breeding population decline
● Breeding range decline
○ Non-breeding range decline

WORDS – JONNIE FISK | **ART** – MIKE LANGMAN

Goldeneye

There are facts. But facts only tell you so much about a bird.

Facts won't tell you what it is like to watch a Goldeneye. And they won't tell you about their introvert's energy, the way that they evade your full attention as they swim, dive, swim, dive. They will tell you that males and females are plumaged differently but not that the male's black-green head shines like velvet in the right light. They say that he bobs that head forward and flings it back in display, but not that it looks like a yoyo while he does so. They'll tell you that males and females tend to winter in different locations. That, at least, is interesting, because this winter there have been six females to two males wintering on this stretch of the River Nith, just beyond the edge of Dumfries. The river is dark here, the water in shade, reflecting bare winter trees. I am stringing together the ripples of the just-dived, the resurfaced-somewhere-else, the gone-while-I-refocused-the-binoculars. Whack-a-Mole has nothing on watch-a-Goldeneye.

When they stop diving, their bold eyes stare back warily as they swim at the speed of my footsteps, drifting towards the far bank, always keeping a fixed distance, unwillingly watched. The drakes fly upriver on whistling wings, leaving the subtly plumaged females. Their head is conker-coloured (and shaped not dissimilarly, that wonky, hand-sketched circle), perched as if without a neck onto a small, soft-grey body that pales where it meets the water. She has a bill that is large and wide for dealing with Crustacea, though what crustaceans they find here I do not know. There is much about Goldeneye that I do not know, hidden as they often are by the dark water.

This is a species that has, and I have searched extensively, almost no cultural history, beyond a habit of nest box use, and a James Bond film that wasn't, I was upset to learn as a child, anything to do with ducks. If you were to reanimate our wildlife, in a distant depleted future, based on our representations, you would probably end up without a Goldeneye. You might not even notice.

Facts make you think you know a species; emotions help you realise you don't. You line up a clear idea of what a Goldeneye is and then it slips away, leaving you clutching at the ripples of the just gone.

CHANGING BOCC STATUS 1–2–3–4–5
RED-LIST CRITERIA APPLICABLE
- IUCN Globally Threatened
- Historical breeding population decline
- Breeding population decline
- Non-breeding population decline
- Breeding range decline
- Non-breeding range decline

WORDS – STEPHEN RUTT | ART – TIM POND

Smew

Last month I went looking for a Smew – the Karl Lagerfeld of diving ducks. Vivid white crest, jet black shades, white tux, moving elegantly in cold clear water. I've never seen one. Numbers of wintering visitors have been tumbling for some time, and the grapevine said there was one, quite suitably, on Tumbling Bay Lake at Amwell Nature Reserve, 45 minutes from me round the M25.

Fossils show birds similar to Smew existed 13 million years ago. And I picture them with the same hook-tipped bill with serrated edge, crisp white outline, probably favouring the same waters – slow-moving, marine intertidal or wetlands. It is most likely 'Smew' is a variation on a Dutch dialect for 'wild duck'. One old English name for it is 'white nun', although the females and youngsters are grey with ruddy, chestnut heads.

Lately, I've launched an environmental campaign to reverse decades of neglect at my local Brent Reservoir (Welsh Harp), where the wetland habitats have been famous for their wintering waterbirds. It's also where Smew have a particular place in the record books. On 28 February 1956, a startling 144 were recorded, a number – by most reckonings – still to be beaten nationally. Smew sightings at the Welsh Harp date back to the mid-19th century, but it wasn't until the 1940s that numbers suddenly took off. For the next 50 years, the bird remained a big winter attraction in north-west London, but by the late 1970s numbers had dwindled to perhaps 10. Today, the last recorded sighting was in 2011, when three redheads passed through.

The reason is almost certainly conditions. Milder winters, the recent 'hottest decade in history', and an increase in watersports, have probably made West Hendon a less attractive destination for the modern-day spartan Smew on vacation. Local water quality probably hasn't helped either. These days, the two inflowing rivers at the Welsh Harp are badly oxygen-depleted, and high in urban run-off, contaminated silt, phosphates and microplastics. Feeding grounds are suffering and the numbers of many regular species are in decline.

The Smew at Amwell Nature Reserve proved elusive. I stayed to watch the wolf moon rise over the lilac water, scope out, but no joy. Odd visitors like this are most likely an outlier from the populations that now gather in notable numbers in the chillier Netherlands. Single waterbodies near Amsterdam can attract 50–200 at a time. Smew will remain a scarce winter visitor in North London now, but I haven't given up hope.

WORDS – BEN WATT | ART – JOHN HATTON

Pochard

Winter has arrived at the fringes of Middlesex suburbia. The air is crisp at my patch, where the Metroland towns of Ruislip, Eastcote, Northwood and Harefield converge. It is nine in the morning and I arrive at Ruislip Woods National Nature Reserve, disembarking from the aged H13 bus. Painstakingly-crafted House Martin nests under the roofs of nearby houses wait for their occupants to return from Africa in a few months' time; and an amorphous and tangled labyrinth of ancient trees creak and whistle in the wind. While the woodland may be quiet and seemingly barren except for the odd call of a Redwing, the former reservoir of Ruislip Lido, at the centre of the nature reserve, is abundant with ducks.

Scattered across the Lido, the ducks wake up to the morning mist. Mallards potter around in the shallow waters near the bench on which I sit, while the equally common but shyer Tufted Ducks dive headfirst beneath the waves, only to pop up again three metres away from where they were a few seconds before. In winter, the plumage of these resident ducks appears muted; the ducks' gorgeous palette of dark greens, mottled browns and Prussian blues is dimmed by the grey of winter. One visitor from Scandinavia, however, catches my eye.

Out of all the ducks on the lake in front of me, the Pochards are truly a sight to behold. The female Pochard appears inquisitive and curious. Gliding with ease across the open water, she turns her head to inspect the feathers of the other wildfowl bobbing up and down past her. She appears to pay great attention to detail, with the white circles around her eyes akin to spectacles. Much of the time she swims directly behind her mate. In contrast to the female, the male proudly sails past the Mallards and Tufted Ducks, vigorously shaking his head. Droplets scatter from his bill, polished black and punctuated with a brushstroke of white. His immaculate feathers are glossy grey, and his russet head shines bright in contrast to the dark blue waves of the Lido. He glares at me authoritatively with his red, beady eyes.

The Pochard is a well-groomed bird that brings joy to an otherwise bleak winter's day and I feel lucky to see these Scandinavian visitors each winter. It saddens me to write that the Pochard has declined significantly in the UK. Eutrophication, poisoning by lead shot and the introduction of invasive species have all taken their toll. It would be terribly unfortunate for such charismatic ducks to become a relic of the past. I cannot bear to imagine witnessing the gradual demise of the Pochard at Ruislip Lido in the winters yet to come. I sincerely hope there is a way to reverse this decline.

WORDS – KABIR KAUL | **ART** – MANUEL SOSA

Scaup

I can count on the digits of two hands and one foot the number of times I've seen a Scaup so, when I was approached about writing this piece, I felt singularly unqualified. But then it occurred to me that my experience was probably typical; the sad fact is that when a bird is on the Red List, relatively few people will have had the privilege of seeing one.

In my part of south Wales Scaup are mostly passage migrants, dropping in for a day, a week, sometimes longer, to rest and feed, before flying on; in the spring to their breeding grounds in the taiga and Arctic tundra, in autumn to overwinter in the shallow seas off the coast of Britain. Occasionally, a bird will overwinter locally, alternating between the more sheltered parts of Cardiff Bay, the reservoirs in north Cardiff, and the waters of Cosmeston Lakes Country Park.

As an Antipodean who's still climbing a steep learning curve, coming to grips with British birds, I have struggled to identify the many members of the duck family, the divers and dabblers. As a general rule of thumb, the males are easier to recognise; with its glossy green-black head and its mottled pale grey back, the male Scaup is not too difficult to pick out amongst the Tufted ducks and Pochards whose company it seeks. But the female ducks, all confusingly brown, can be a nightmare. And that is undoubtedly why a female Scaup was the star of one of my proudest birding moments.

Just outside the little bay by St David's Hotel in Cardiff Bay, three ducks were cruising, quite a distance away, with the sun low in the sky behind them. When I looked at them through my bins, I saw they were a pair of Tufted Ducks and, amazingly, a female Scaup. She was a little bigger in the body than the Tufties, her head shape was different, and she didn't have their characteristic head tuft or the turned-up Tufty tail. She had quite a large white patch on her face, but some female Tufties also have this, and she looked greyer in the lower body, though that could've been a trick of the light.

I belong to a local WhatsApp birding group, though I hesitate to post sightings unless I'm sure of what I'm seeing. This time I was brave and posted a message, with a big 'maybe' proviso. Another of our group, an expert birder with many years' experience, happened to be on the shore opposite, so he looked for the birds from his vantage point. And he agreed with me. It was an unforgettable moment – I had found my very first female Scaup!

WORDS – ANNIE IRVING | **ART –** DAVID BENNETT

Red-necked Grebe

I saw my first Red-necked Grebe in July 1994 at Dosthill Quarry near Tamworth. The summer of '94 was significant for British birders because there was a huge influx of Ruddy Shelducks, which is why I was at Dosthill. We'd just dropped my nan off at Birmingham Airport for an early flight, so on the way home to Stoke I persuaded my parents to take a detour to see a couple of these, which I seem to remember had been present for about a week. My parents were used to leaving me in strange places for a couple of hours, so off I went with a can of Lilt and a Double Decker, whilst they went to find a Little Chef for an Olympic Breakfast.

I found the two Ruddy Shelducks within a few minutes; they were nice enough, but what amazed me was the swarm of Great Crested Grebes – my notebook says at least 100. It was already a hot morning, so I lowered my tripod and sat on the banking, trying to find anything else other than a Great Crested Grebe. Suddenly within the throng, up popped a Red-necked Grebe, the first I'd ever seen, but no messing around – it was a gorgeous full summer adult. And then it was gone. I knew this was a really good local bird, but I couldn't re-find it. For another half hour I panned my telescope from side to side with no luck and started to panic that I'd imagined it.

I could see two other birders coming along the path and so I needed to make a decision – tell them about the grebe and hope they'd pick it up, or forget the whole thing and put it down to a sugar-induced hallucination from Lilt and Double Decker. Nowadays I couldn't care less, but when I was younger, the idea of messing up birds or being accused of an over-active imagination would have murdered me with embarrassment. It's ridiculous that birding can get to you like this, but I was genuinely worried about how this might play out and potentially ruin my entire life as a birder. So, as they approached, I opted to bail out and not mention the grebe.

Their first words to me were: "Have you seen the Red-necked Grebe?" Massive panic over – it transpired that it was found late the night before. I didn't manage to pick up the grebe again before my parents came back to pick me up; and they didn't find a Little Chef either.

WORDS – TOM McKINNEY | ART – NORA LIGUS

Slavonian Grebe

There are two roads from the town to the Highland village: the old and the new, on either side of the loch. The new is a scar that cuts across the noses of the hills, forcing its two lanes and white lines where no road was ever intended to be. It screams with timber lorries and mobile homes and Germans in the wrong lane. In summer a nightmare.

The old road meanders along the ridge on the edge of the Monadhliath Mountains the other side of the loch and the Great Glen fault, following the line of those who made it with pick and shovel 200 years ago. It wanders high above the valley, curling round rocks and over humps and narrow bridges. Sometimes it just stops, takes a deep breath and realigns. Few use it except locals, for in winter it is treacherous. It arrives above the village and tumbles down a staircase of twists and bends to reach the lower ground. In one place it forms the edge of a small lochan of crystal water in a scoop of rock high above its destination. In the distance the hills above the fort 30 miles away are hazy with blue cloud-smoke. It was here, one late spring, that I saw a dash of iridescent gold and red on the grey water: Slavonian Grebes.

That was 40 years ago and never forgotten, because it came at the end of a week that seared itself into my brain. Persistent high pressure gave a run of fine weather loved by tourists but hated by foresters such as me. My 50,000 acres of forest and open moorland in the Great Glen were tinder dry, the dead *Molinia* not yet displaced by fresh, juicy new shoots. During my week on duty we had two fires, both on top of 1,000 ft hills and both at night. At the end of it, I went fishing up in the Monadhliath, using the old road. Returning home empty-handed, I passed one of the few flat areas where the road crosses land with a toupee of *Vaccinium*, *Eriophorum* and *Calluna* favoured by Roe Deer and Black Grouse: a Scottish tundra. This is close to where the grebes set up home, and where that brilliant, brief flash of gold and red burned a hole in my exhausted, smoke-filled frame.

CHANGING BOCC STATUS 1–2–3–4–5
RED-LIST CRITERIA APPLICABLE
- IUCN Globally Threatened
- Historical breeding population decline
- Breeding population decline
- Non-breeding population decline
- Breeding range decline
- Non-breeding range decline

WORDS – JIM PRATT | ART – PAUL HARFLEET

Turtle Dove

The purring of the Turtle Dove is in our cultural DNA. Its soft, melancholic "*turr-turr*" once soothed the ears of Chaucer, Spenser and Shakespeare; its pair-bonding an allegory of marital tenderness and devotion, its song a lament of love lost. But the 'turtle' – as the poets called it – is vanishing into the realm of phoenixes and unicorns. At Christmas, when we sing of the gifts my true love gave to me, few carollers, breathing the wraith of its name into the night air, will ever have been stirred by its song, let alone remember one in flight.

So, when I see one, in our rewilding project at Knepp, territorially perched on the withered limb of an oak tree, exactly as Shakespeare described it, I feel the burden of borrowed time. I must score every detail into memory. I watch the effort the dove makes to throw its burbling purr – feathers ruffled, dusky-pink throat inflating, the downward beat of its tail, beak arhythmically opening – and the distinctive black-and-white-striped patch on its neck reminiscent, it seems to me, of the zebras it would have flown over on its migration from sub-Saharan Africa just weeks ago.

When I was growing up in the 1960s, this was the sound of lazy summers, as comforting to me as the call of the Cuckoo. There were quarter of a million Turtle Doves in the UK, then. Now there's just a few thousand. A 2021 survey of Sussex, one of its last redoubts, recorded just 80 singing males. A quarter of these were at Knepp.

Knepp may be the only place in the UK where Turtle Doves are rising and we don't know precisely why. Once present in flocks of hundreds, they're now so few and so shy it's almost impossible to study them. We haven't observed them feeding here or been able to ring them. But most likely they're benefitting, like Nightingales and so many smaller songbirds, from the resurgence of thorny scrub – a safe place to nest. And from an abundance of food sources – for the Turtle Dove, the tiny, protein-rich seeds of wildflowers like Fumitory, Scarlet Pimpernel, Round-leaved Fluellen, vetches and vetchlings, and possibly small-shelled snails. And clean water – so rare, today, in our polluted land.

It's with almost unbearable sadness that I let my heart fill with the Turtle Dove's lament. It trails with it the loss of so much more. Yet can I also hear in it an undertone of hope, perhaps a gentle pleading to endure? Next spring, we may find another territory or two at Knepp and, slowly, as the will to recover nature gathers momentum, its song will be a lament no more.

WORDS – ISABELLA TREE | ART – BEATRICE FORSHALL

Swift

Some years back, in one of of those bleak English springs when every southern migrant seemed to be held up, I went on a miserable search for news of Swifts online. I wasn't alone. Messages were flashing across Europe. "Parties flying through the Pyrenean passes." "Be with you soon in the UK". It felt like plugging into a Swift-dependence support group, anxious souls who'd glimpsed the implications of Ted Hughes' immortal lines in *Swifts*: "*They've made it again,/ Which means the globe's still working ...*" What would it mean if they failed to make it back?

Swifts have been my talisman since I was a kid. I used to walk to school on May Day holding my blazer collar for luck, willing my first bird to appear on this special festive date. A few days later I'd stand in the garden, gaze west over the town and count to 10. If a Swift appeared it meant that they were not just back, but back in the parish. What did they mean to me then? Partly a reassurance that the summer was settled. But also that I was settled too. Swifts are intensely tribal, and their neighbourly packs patrolled the same quarters as our gangs. For a lot of my life, one of the best parts of summer was toasting their pack races round our favourite canalside pub, blissfully called The Rising Sun. This tribe nested in a nearby street of Victorian terraces, and on warm evenings they strafed the cut, past out raised pints, then split apart like a star-burst over the adjoining factory. It was impossible sometimes not to cheer, and feel an empathy with this intoxicating display of full-pelt aerobatics and communal joy-riding. They were a clenched fist raised against entropy. Years later I was watching a live news bulletin of the shelling of Beirut. Halfway through the reporter was suddenly surrounded by a flickering ectoplasm of Swifts, silhouetted against the flashes of exploding shells. "Choose life" they seemed to scream.

Now when I read that their population in the UK is already down by 50 per cent I wonder what it would mean to lose them. They come closer to us than almost any other bird. They nest inside our houses. On the banks of Tring Reservoirs and in the back-alleys of Granada, they've flown so close that I've felt the rush of their wings on my face. But not for one moment do Swifts ever seem to be cosily part of our world. I once glimpsed a pair mating high in the air, and for a moment they looked like a four-winged creature from a bestiary. Swifts are a vital portal between the wild world and our own.

CHANGING BOCC STATUS 1 2 3 4 5
RED-LIST CRITERIA APPLICABLE
- IUCN Globally Threatened
- Historical breeding population decline
- Breeding population decline
- Non-breeding population decline
- Breeding range decline
- Non-breeding range decline

WORDS – RICHARD MABEY | **ART –** HARRIET MEAD

Cuckoo

*"April come she will,
May she's here to stay"*

It's 28th May, 3 am and pitch dark. I step out of my car in front of a high fence, a sign saying, Fylingdales Early Warning Station – KEEP OUT. I am excited. I watch my breath condensing in the air. Soon I see approaching headlights. Out jumps Chris Hewson and his crew from the British Trust for Ornithology. They are here to catch Cuckoos, trying to solve the puzzle of their disappearance.

We are escorted onto the site through the rolls of razor wire. We have our mugshots taken, passports examined. It never used to be like this to see a Cuckoo. Eventually we're bumping across the moor and as we reach the great monolith with its heaven-facing discs, I hear your voice reaching out through the grey light of dawn – "CUCKOO". I trill inside, meeting an old mysterious friend.

When I was a child, you were so familiar. Even in suburban London you announced the spring as reliably as the sun would rise. Now, I feel a deep yearning when I hear you. Your voice speaks of loss, a hole in the world where you once were. But you have always been elusive, Cuckoo. Invisible one, constantly disappearing, a flash of wing and you are gone, throwing your voice to someone else on another hilltop.

You suddenly dart past low over the heather, pursued by a flock of pipits, wings swept back scimitar-shaped. You take my breath away. There is something otherworldly, shamanic about you, your golden eye holding an untouchable wildness, challenging me to cross into a nameless dimension.

We set up the mist nets and goad you towards us with a recording of a female's bubbling call. A group of us: photographer, journalist, military man, scientist and storyteller move to a heathery mound and wait. Soon the net bulges and Chris retrieves you, bringing you back to the Land Rover in a white cloth bag. You are measured and a satellite tag attached to your back. Your yellow eyes stare defiantly. Mick, the fierce bird-loving Yorkshireman who brought us here, holds you in his hands and throws you in the air as if he was releasing his own spirit. You disappear, into the mist of dawn, as Cuckoos always do.

*"June she changes her tune,
July she prepares to fly,
August away she must."*

CHANGING BOCC STATUS 1 2 3 4 5

RED-LIST CRITERIA APPLICABLE
- IUCN Globally Threatened
- Historical breeding population decline
- Breeding population decline
- Non-breeding population decline
- Breeding range decline
- Non-breeding range decline

WORDS – MALCOLM GREEN | **ART –** STEVE CALE

Corncrake

Fresh off the ferry from Mallaig, we tramp towards Eigg tea-room, with our rucksacks and instruments, ten folk musicians arriving for a residency called Songs of Separation. I stop at a wildlife interpretation board, and I'm struck by an illustrated Corncrake. The Small Isles are amongst the few places in the UK and Ireland where you might encounter one of these reclusive birds. From my interior jukebox, a melody rises –

"wi her I spent some happy nights, whaur yon wee burnie rowes
and the echo mocks the Corncrake, among the whinny knows"

It's a 19th century love song from south-west Scotland. As I haul it from memory, there's a sadness in it, which hasn't occurred to me before. Both the song's narrator and its intended listener know the Corncrake intimately. Indeed, its abrasive call – "*crex crex*" (the bird's onomatopoeic Latinate name) – is synonymous, in the song, with summer itself.

Two hundred years ago, Corncrakes could be heard from Edinburgh's Charlotte Square, over fields where there's now a sea of Georgian townhouses. There are no longer any Corncrakes at all in southern Scotland, where I live. During the late 20th century, they were pushed north and west, to Argyll and the Hebrides, Orkney and Donegal, by agricultural mechanisation and intensive farming. To know they're here on Eigg is tantalising.

I bring *Echo Mocks the Corncrake* to Songs of Separation, as a trace of a time when we humans, at least in these parts, were less separate from, and in opposition to, the life around us. There's no time to seek out an actual Corncrake. Instead, fiddle and viola pluck pizzicato *crex crex*, a bass bow creaks, fingers drum a guitar's body in mimesis, and I sing –

"rural joy is free for a', whaur the scented clover grows
and the echo mocks the corncrake, among the whinny knowes"

A Facebook message arrives, from a former colleague. *I hear you're on Eigg singing about Corncrakes*, she writes. Attached is a blustery iPhone recording from just over the hill towards Laig: "*crex crex*". Maybe next time I'll head inland. 'Til then, there's song –

"The Corncrake is noo awa, the burn is tae the brim
The whinny knowes are clad wi snaw, that taps the highest whin
when cauld winter is awa, and summer clears the sky
we'll welcome back the Corncrake, the bird o' rural joy"

WORDS – KARINE POLWART | **ART** – SARA RHYS

Leach's Storm Petrel

European Storm Petrels always seem to be in a hurry, their bat-like flight scudding them across the waves in a rush to the next feeding opportunity. By contrast, Wilson's Storm Petrels like long glides and dangling their feet, but for me the ultimate storm petrel is Leach's – once seen, never forgotten.

In the UK, Leach's Storm Petrel doesn't come easy, but I have been lucky enough to connect with this truly pelagic seabird on several occasions and every single one has been special. I have seen them easing themselves over mountainous wave tops in the teeth of a ferocious storm, on far flung islands in the middle of the night and from the gently wallowing deck of the Lily of Laguna, fondly known as the green submarine, in the south-western approaches.

The Lily is special too; she is one of those boats that is full of character, her open deck giving one the full, immersive experience of pelagic birding and, as one of the small boats that joined the dash across the Channel to help evacuate troops from Dunkirk, it is always an honour to be birding from her.

Sea-watching from a small boat is a very different experience to doing the same thing from land, the biggest difference being the proximity to the birds – some pass by within a few metres and, as such, the views can be stunning. It was under these conditions that I enjoyed my best ever Leach's Petrel.

The light, graceful, butterfly-like flight instantly set this bird apart from the European Storm Petrels that were also feeding on the chum put out by the skipper, but it was the silky dark-milk chocolate plumage that stunned. With a shining white rump, split down the centre with a dusky smudge, and a shallow forked tail, it seemed to me to be the most beautiful Leach's Petrel ever. It was as if I was seeing one for the first time.

I spent seven years living on the Isles of Scilly with my family, our daily lives lived cheek-by-jowl with the sea, 30 miles south of Land's End. During this time I saw thousands of European Storm Petrels, over 30 Wilson's Petrels but only three Leach's. Most of these birds were seen from small boats around 10 miles south of St Mary's but only one of the Leach's Petrels was seen in this way; the one described above. It really was a very special bird indeed.

CHANGING BOCC STATUS 1 2 3 4 **5**
RED-LIST CRITERIA APPLICABLE
- IUCN Globally Threatened
- Historical breeding population decline
- Breeding population decline
- Non-breeding population decline
- Breeding range decline
- Non-breeding range decline

WORDS – PAUL STANCLIFFE | ART – REN HATHWAY

Balearic Shearwater

One of the wonders of birding is watching the effortless command birds have over the sky. Just as we might blink or scratch our noses, birds glide, shear, dive and manoeuvre in the air with such ease we can't help but wonder at them. Arguably the champions of them all are most at home far out to sea, all the better for strong winds and high waves; the shearwaters.

The Balearic Shearwater has a breeding range limited to the coastal cliffs of the Balearics, where these birds face the threats of climate chaos and the local pressures that come from the development of holiday resorts and introduction of non-native mammals. At the end of the breeding season they do something unusual, migrating out of the Mediterranean and north into the Bay of Biscay, where they moult, before then heading south to winter off West Africa. During late summer they may be encountered off the UK's southern shores and, if the winds are right, along North Sea coasts.

Of all the peculiar niches within birding, and without doubt the strangest and most dedicated, is the cult of sea-watching. There are many crosses to bear as a birder, and never more so than as a sea-watcher. Hours pass, day-after-day. The best spots are the most exposed, and the best weather the worst you can handle. Staring at the sea, ever-changing but consistently the same. A rugged headland, swirling swell below. A cutting northerly wind bringing frequent stinging rain showers, and there on a camping chair with a flask of tea, a huddled figure, trying to keep the notebook dry and the scope steady. You would be excused to think this is madness. But consider, this is where these birds flourish, in a place so inhospitable to us we shake our heads and turn away for home. Where the sea forces us to stop and back away, the shearwaters thrive.

The regular trickle of Manxies has propped up many a sea-watch. You see that familiar black-white flash but then something different: a dark smudge, diffuse and lacking contrast, a different outline, it's obscured behind a wave and then, it looks pot-bellied and heavy, another Manxie? … no, a Balearic Shearwater. The Mediterranean visitor shearing low to the sea has just entered your world. You watch it glide north completely unaware of the spike in adrenaline it has just caused. Desperately you try to drink it in, making a note of features and flight pattern and marvelling at one of the rarest seabirds crossing your view. Think for a minute of the path this bird has taken to reach you. Critically Endangered, one of approximately 19,000 left and living a fragile existence in our changing world. Extinction looming in the next 60 years. And there it goes, out of sight behind a wave. You can start to breathe now!

WORDS – DAVID STEEL | **ART** – OLIVIER LEGER

Shag

As a boy, I had learned Christopher Isherwood's poem, *The Common Cormorant (or shag)* – with its strange mix of paper bags, bears, buns, and lightning – by heart, and long before I had first set sight on the bird itself. But a move from Buckinghamshire to Cornwall turned this mythical animal into a near daily reality, as Shags are regular along the north coast.

Close to home, one small cove supports half a dozen pairs each year, and I loved lying in the warm heather watching their antics, not least the gathering (and robbing) of seaweed to build their nests. Shags strike an ancient cruciform pose in flight, and when 'drying' their wings. In the early days, I watched them from afar with my first pair of binoculars, their awkwardness on land a contrast to their evident delight in the surrounding seawater.

From a distance, Shags appear black, and almost reptilian, gaining distinctive forward-curving crests in the breeding season, possibly a source of one Cornish name, 'shagga'. In those early days, I had Peterson's field guide, and 'The Cormorants' did not justify a plate, but were drawn out on a page in simple black and white. Later, having secured a kayak, I started to explore the cliffs from the sea edge, and discovered the Shags' striking colours. Another early name is Green Cormorant, and adult birds are indeed a glossed deep-green, tinged violet and scaled black at close quarters. With yellow gapes and near-emerald eyes, and surprisingly at ease with a quiet paddled approach, they were astonishing.

Summer swimming in the harbour included – predictably – competitive breath holding, not that we could out-dive a Shag. They swim down to an average depth of 30 metres for up to over a minute. In William Yarrell's *British Birds* – published in 1843 – he noted "one had been caught in crab-pot fixed at twenty fathoms", and they are a regular sight for the artisanal lobster-potters that venture out of Boscastle when the weather allows.

In deep winter, western storms and swells make the coast a dangerous place. Boats are stored ashore, and the local gig club takes to training on inland waters. Writing in 1829, the great Cornish naturalist, Jonathan Couch, noted that both Shags and Cormorants were common, and that they "become of a plain Colour in the winter"; so echoing the bluff headlands where the salt wind subdues the heathers, Blackthorn, and Gorse to brown. Skeletal single Hawthorn trees stand out, echoing the angular shapes of Shags below.

WORDS – JOHN FANSHAWE | **ART** – DANIEL COLE

Dotterel

Dot, dot, dot, Dotterel.

Do you dash? Cocoa dots, marmalade cuff.

Dot, dot, dot, Dotterel.

Can you dance? Leaves fold – hide, hide.

Dot, dot, dot, Dotterel.

Disappearing – a dash,

dashing off the page.

CHANGING BOCC STATUS 1 – 2 – 3 – 4 – 5

RED-LIST CRITERIA APPLICABLE
- IUCN Globally Threatened
- Historical breeding population decline
- Breeding population decline
- Non-breeding population decline
- Breeding range decline
- Non-breeding range decline

WORDS – ALICIA HAYDEN | **ART –** MARK GURNEY

Ringed Plover

The gentle scuttle over the rocks, the unmistakable bum-bob and the elegant glide into a nest site – it can only mean one bird: the Ringed Plover.

I've loved the Ringed Plover since I was a girl; I had one painted on my bedroom wall. They were in abundance, and everywhere we went I could hear the gentle call of "*poo-eep, peep-peep-peep*" of the nesting birds on our local estuary. Even a trip to the local beach, sporting my finest cargo trousers as a child, was never complete without watching the Ringed Plovers scuttling over the pebbles on the shoreline, chasing the food on the incoming tide.

Now, as an adult, they've played a part in my life and career. From stays at Bardsey Bird Observatory, where I would spend hours laying among the sand watching the birds on the shore, to my time volunteering with the Little Tern colony at Spurn Bird Observatory, I would spend my days mesmerised by the alarm-displays, that incoming glide they so often do around their nests, and the eerie "*poo-eep*" calls in the distance.

They're a species that has always entertained me. Watching them doing their bum-bobs in the sand, and their ever-funny way of searching for food – running then stopping with their heads barely above the neighbouring pebbles on the shore – nothing quite captures the essence of childlike behaviour like a Ringed Plover. Their breeding style has always amazed me too. The sheer determination to defend their eggs by screaming and running has always been a puzzle when you compare it to the aggressive style of Oystercatcher and Avocet, with which they often share their nesting beaches.

Over the years my love for the species has grown, as they continue to follow me around on my journeys and career; even abroad I continue to spot the species. It still seems that Ringed Plovers are doing well but it's only when I think back to my childhood, and how many I would see on our estuary outside of the house, that I can see the change. Now, when I visit the estuary I barely see a Ringed Plover let alone a breeding pair. The love continues to grow for our delightful Ringed Plover, even if the population of the species does not.

WORDS – DAN ROUSE | **ART** – ROSIE VILLIERS-STUART

Lapwing

I heard them before I saw them. Like the plaintive squeak of a rubber toy, or an oboist tuning up, it was unlike anything I'd heard before in the Brede valley. Then, a sudden explosion upwards of a bird with black-and-white wings as long and elegantly curved as an owl's, pirouetting away from the reedbed and plunging back down – then up, up again, now joined by another. Like a matador's cape, those wings whoomphed up and down, switching this way and that, graceful and athletic and wholly surprising.

An older man with binoculars stopped to watch. What were they? "Green plover, peewit, Lapwing – take your pick. They're chasing the Rooks off their eggs." He lent me his binoculars, and I spotted a quietly grazing bird with a head crest like an upward stroke of calligraphy, its folded wings now the greenish hue of oil-on-water. We were, he said, lucky to find them. "See that field?" – a dark green expanse next to the ribbon of pasture and reedbed, stretching east to west. "Wheat, planted in November. Used to be planted in the spring. Lapwings can't breed in a mature wheat field." Had the landscape much changed? He smiled ruefully. "It was like a patchwork quilt, when I was a boy. You'd see 500 pairs of Lapwing on a field in winter, maybe more."

And not just in Sussex. Lapwings are embedded in place names throughout Britain: Pyewipe near Grimsby, Twitfield near Lancaster, Tivetshall St Mary in Norfolk. 'Lapwing' comes from an old English word, meaning 'leap with a flicker'. Its vernacular names are even better. Horneywink. Teewhuppo. Toppyup. Chewit. Peasiewheep. Tuefit. Tiecks nicket – or, in Shetland, tieves nacket: 'thieves' imp'. Such playful, humanising names hint at a long relationship with man. A 'deceit' of Lapwings, the collective noun, refers to their sly theatrical skills in distracting predators away from their eggs. And it was the eggs we humans once wanted. Prized as a delicacy, served in baskets lined with moss to emphasise their speckled beauty, Lapwing eggs were plundered well into the 20th century to feed the London market.

This bird, so interwoven with our countryside, language, habits and rhythms, was once an agricultural totem. My *Observer's Book of Birds*, 1972 edition, sites the Lapwing in "large numbers" on ploughed land, pastures and moors, especially in the winter. It is "one of our most useful birds" thanks to its appetite for snails, slugs, and "all kinds of injurious insects and their larvae". Since then, with the intensification of farming, some 80 per cent of these shape-shifting imps have disappeared from southern England and Wales; 50 per cent in the past decade alone.

WORDS – TESSA BOASE | **ART –** JO WRIGHT

CHANGING BOCC STATUS 1–2–3–4–5
RED-LIST CRITERIA APPLICABLE
- IUCN Globally Threatened
- Historical breeding population decline
- Breeding population decline
- Non-breeding population decline
- Breeding range decline
- Non-breeding range decline

Whimbrel

On a clear day, looking east from the Island of Lundy, you can see up the Bristol Channel; Pembrokeshire and the Gower sit to the north and the North Devon and Cornish coast stretches back to the south. The sight makes you think of the tiny island's place in the country, and world, as thousands of birds pass over on migration. In spring that includes many Whimbrel, birds that take the opportunity to feed in the fields or have a brief rest on the tavern roof before heading to their northerly breeding grounds. Whimbrel have been doing this journey, between tundra and heath of northern Europe to mudflats and mangroves of Africa for at least 1.9 million years.

The Whimbrel's call is cloaked in folklore. It has a reputation because of its haunting quality, linked to the fabled seven whistlers and believed by some to be the call of the dead returning to warn the living of impending doom. Others believe that when a group of six is joined by a seventh the world will end, the subject of the Iron Maiden song, 'The Prophecy'. Icelanders, however, consider the song a little more mundanely, its notes like a pot of boiling porridge.

An increasingly precious few Whimbrel head to Scotland, mainly Shetland, to breed. The UK breeding population has varied; it once looked so promising, jumping from and estimated 150 pairs in the 1960s to 470 pairs in the late 1980s; but the latest population estimate now stands at 290 pairs.

After their impressive mating displays, Whimbrel settle down for the breeding season in monogamous, doting, protective pairs. Chicks are born downy and raring to go, leaving the nest within a few hours. Despite such a specialised beak, Whimbrel are not fussy eaters, taking crabs, sea cucumbers, berries, and grasshoppers; it seems that no phylum is safe!

Living on Lundy can lull you into a false sense of security in terms of bird numbers. Our location attracts the passage of many rare species, needing to reorient, snack or rest before completing their journeys. In the 1960s, counts of over 100 Whimbrel in a day were recorded, probably reflecting the former importance of the Somerset and Gwent levels stop-over which, now drained of boggy pasture, no longer supports the thousands of migrating Whimbrel it once did. Although sadly much reduced, these birds provide the soundtrack of island spring migration, with all admiring their gangly elegance and hoping recent conservation efforts will halt further decline.

WORDS – ROSIE ELLIS | **ART** – JIM MOIR

Curlew

What does it mean then, to sing? To venture a note upon the air; to feel your body and your mouth resound with this resonance of flesh, and mind, and heart. To sing, to really sing, requires both complete focus and a searing commitment to being present within the passing moment. To be responsive to its signals, and open to its commands, that the voice might somehow channel something – something raw and elemental, something true, something from the gut – and in doing so, begin to tell a tiny fragment of the story. The story of how it feels, deep down, to be a part of this vibrant, sonorous world. To be tied by blood and breath to every living thing. To be wind-blown and alive.

If I were to describe the Curlew, should I describe the bird that I see? The dainty stepper, the keeper of wind-ruffled silences. Mottled mud spirit, keen-eyed sky spirit … This strange, thoughtful creature probing its own reflection, dressed understatedly in streaked browns and greys, with small head, long legs and neck, and huge, new moon slither of a bill.

Or should I describe the bird that I hear? The haunting sound that issues forth when that great bill opens. A song that seems to split in mid air and take the heart in two directions at once. An unforgettable, transcendent beam, which as it rises, seems also to send down a great tap root into the deepest shadows of the land's memory.

Neither description can bring you close to knowing the bird as it truly exists – the Curlew as the shape and voice of wildness itself; as essence of place. Encountered in the wild, its mysterious presence and eerie music provides a kind of mystical stitching that not only binds the landscape together, but also fastens present to past. To hear a Curlew's song as it carries across the land, is to hear, quite literally, a place being sung into being. By extension then, to lose that song would be no less than losing a part of the land itself, and by further extension, a part of ourselves.

What kind of nation would allow such an ancient magic as the Curlew to be marginalised and destroyed? If we are to turn the quickening tide, we must shake the white noise from our heads and attune ourselves once more to the Curlews' plaintive cries. These birds are kindred spirits, and these words a small acknowledgement of the debt of one singer to another.

An unsung land is a dead land.

WORDS – DAVID GRAY | ART – MELANIE MASCARENHAS

Black-tailed Godwit

Over the last three decades, I have been fortunate to become extremely familiar with Black-tailed Godwits, which are surely among the most elegant and enchanting of wading birds. Together with colleagues and birdwatchers from across Europe, we have tracked the journeys of individual Godwits, sometimes throughout their lives of more than 20 years. These birds become old friends and open our eyes to the world of migration.

During this period, the fortunes of different populations of Black-tailed Godwits have varied greatly, with plummeting numbers of the *limosa* population, that breeds in western Europe and winters from southern Europe to west Africa, while the *islandica* population that breeds in Iceland and winters in western Europe has continued its century-long population expansion.

These contrasting fortunes reflect differing changes to breeding conditions. The extensive meadows in which *limosa* Godwits primarily breed have largely been converted to intensive grassland and arable production, reducing nesting opportunities and increasing losses of eggs and chicks to predators and mechanical agricultural operations. By contrast, the semi-natural grasslands of lowland Iceland that support *islandica* Godwits have experienced increasingly benign climatic conditions, fuelling high levels of productivity and population growth. In the UK, numbers of wintering *islandica* Godwits have flourished while breeding *limosa* Godwits are now so rare that it has been necessary to invest in headstarting to try to prevent their local extinction.

Both *limosa* and *islandica* populations have been the subject of decades-long tracking studies which have given us extraordinary insights into the life-long experiences of individual birds – the trials and tribulations they face in attempting to raise chicks, and the perils of migration. We have learned that individuals are remarkably consistent in their spring migration timing and that each individual visits the same few sites, repeatedly, throughout their lives. Migratory birds have extraordinary capacities to navigate consistently in space and time, yet individuals are extremely vulnerable to the loss of the particular sites on which they depend.

While burgeoning flocks of wintering *islandica* Godwits on UK estuaries have been a wonderful sight, there are concerns that they too may soon be impacted by environmental changes. Widespread drainage and planting of non-native forests in lowland Iceland are breaking up the open tracts of wetland that support many breeding waders. Ensuring that waders have the space they need will be critical if the spectacles of their breeding displays and vast winter flocks are to be enjoyed by future generations.

CHANGING BOCC STATUS 1 • 2 • **3** • 4 • 5

RED-LIST CRITERIA APPLICABLE
- IUCN Globally Threatened
- Historical breeding population decline
- Breeding population decline
- Non-breeding population decline
- Breeding range decline
- Non-breeding range decline

WORDS – JENNIFER GILL | **ART** – NIK POLLARD

Ruff

Wader identification is tough,
      a dash of brown, a flash of buff,
A little round head – and sure enough –
      I find all these (and more) on the Ruff.

With its medium legs and medium size,
      medium bill and small dark eyes,
The Ruff is a master of disguise –
      I fail again. It's no surprise.

So woe betide the ornithophile,
      that seeks a Ruff while juvenile,
The search may last for quite a while –
      and many others share its style.

But then, good grief, in May and June,
      I'm treated to a different plume,
A glorious crown, I gush and swoon –
      and then, no doubt, I change my tune.

"The Ruff!", I cry, "in all its splendour!",
      "Clear as day, that's no pretender!",
Now I know (and must remember) –
      just seek out the ZZ gender

But then, alas, once all have bred,
      All colours gone, all feathers shed,
I'll stare at the marsh with familiar dread –
      peer through bins ...
            ... and scratch my head.

WORDS – JESS FRENCH | **ART –** ANNA TERREROS-MARTIN

Dunlin

In late August 1976, the Wash Wader Ringing Group caught and ringed 3,541 Dunlin on The Wash, a vast rectangular estuary on England's east coast. It was an amazing week yet, at the end of the trip, most of the birds being trapped were unringed; who knows how many Dunlin there were altogether? For those of us who were there, memories of sunny days, swirling flocks of waders and blisters from closing metal rings still linger. How can a bird that was so numerous less than 50 years ago now be Red-listed?

In the mid-seventies, the Dunlin was the most numerous wader on the Wash, with peak Wetland Bird Survey counts of up to 70,000 individuals. Now, Dunlin has slipped into third place, with maximum counts of only 26,000. These totals are only part of the story; they don't account for the turnover of birds over the course of a year.

The first Dunlin to reach The Wash in July are of the *schinzii* race, most having flown from southern Scandinavia and the Baltic. Many of these early birds, with their orangey-brown summer plumage, fatten up before heading on to Africa. *Alpina* race birds, with chestnut colours in their backs and blacker tums, arrive on The Wash later from further north and east. These larger birds will moult their feathers here and stay longer, some for the whole winter.

Across the UK, counts of winter *alpina* Dunlin are only 40% of those in 1975. There is a theory that, as winter conditions on the continental side of the North Sea have become less harsh, new generations of juvenile *alpina* have settled in countries such as the Netherlands, instead of continuing their south-westward migrations from northern Russia. Individuals are site-faithful so, once a wintering site has been chosen, one ends up with newer generations on the Continent and proportionately more ageing adults in the UK and Ireland.

The inclusion of Dunlin in the UK Red List is justified by declining winter numbers alone but there is probably more concern about *schinzii*, the race that breeds in the UK. It's hard to work out how rapidly numbers are falling but alarm bells are being sounded in Ireland, where there are fewer than 50 pairs, and in the Baltic, where numbers have fallen by 80% since the 1980s. Poor annual survival rates suggest that *schinzii* Dunlin face problems in West Africa or on migration so we need to do as much as possible to support the UK's breeding pairs.

WORDS – GRAHAM APPLETON | ART – CELIA SMITH

Purple Sandpiper

It's an easy species to overlook, the Purple Sandpiper. For a start, they're not purple. Yes, in strong sunlight and at the right angle, you can catch a slight sheen off the feathers on the back, but for all intents and purposes, they're grey. They spend their time creeping over similarly-coloured grey rocks, like mice – indeed, one of the old Northumbrian names for them was 'sea-mice'. They are rarely found in large numbers, and often occur in the company of louder and more eye-catching Turnstones; as I said, easy to overlook.

In the UK, 'Purps' are best found by heading to northern coasts – particularly in north-east England and Scotland – and searching on rocky shores, right on the edge where the waves crash in. You'll be rewarded by finding a quietly charismatic species; beady eyed, with orange legs and bill base, and confiding, happy to feed close by if you stay quiet and still. You'll be able to marvel at the toughness of these little birds as they skilfully dodge the breakers. They breed on tundra habitats as far north as Franz Josef Land, at 81°N, and many spend the winter on Arctic coastlines, further north than any other wader. But every winter about 10,000 (at the latest count) leave breeding grounds that stretch from Arctic Canada to Svalbard, seeking the comparatively milder climes of the UK.

Sadly, numbers wintering here are dwindling: down by 34% between the winters of 1997/98 and 2015/16. This might be a consequence of climate change, with warming winters meaning fewer birds feel the need to migrate south. However, it may also indicate a genuine decline in the population. The recently published *European Breeding Bird Atlas 2* shows a widespread contraction in the breeding range in Scandinavia since the 1980s, a pattern reflected in several northern breeding wader species.

The Red-listing of Purple Sandpiper is, however, due to a decline in the UK's tiny breeding population, found in mountainous, tundra-like areas of northern Scotland and sporting their brown and spotty breeding plumage (still not purple!). Only a single returning pair has been reported to the Rare Breeding Birds Panel in recent years. The fall in numbers may simply be the sort of fluctuation that tiny, vulnerable populations often undergo, but with other montane species such as Dotterel and now Ptarmigan also on the Red List, it could reflect wider pressures on these habitats which, again, might include climate change.

CHANGING BOCC STATUS 1–2–3–4–**5**

RED-LIST CRITERIA APPLICABLE
- IUCN Globally Threatened
- Historical breeding population decline
- **Breeding population decline**
- Non-breeding population decline
- Breeding range decline
- Non-breeding range decline

WORDS – MARK EATON | ART – RUTH WALKER

Woodcock

Woodcock moon; for many, the Woodcock's return to the UK coincides with a change in season. Whilst most individuals are resident throughout the year, there are hundreds of thousands which fly 3,000 km from the frozen landscapes of Russia and Finland to our lowland habitats to overwinter in the UK's milder climate. These standard facts are ones we take for granted today, but once upon a time there was much speculation and confusion about the pattern of Woodcock movements.

The Woodcock's nocturnal habits, secretive behaviour and cryptic camouflage make this a challenging bird to see, and these features have cloaked this unusual wader in fascinating folklore and mystery for centuries. Marvelling at migration, it was once a commonplace belief that they'd fly to and from the moon during the months they were absent from the UK and would often carry smaller birds on their back, who were assumed not to be physically able to make the long journey alone (famously the UK's smallest bird weighing only 5 g – the Goldcrest).

After the Woodcocks' true flight paths to northern Europe had been identified, it still puzzled scientists how they managed to fly such a long distance over hostile open water. The history books depict tales of Woodcock carrying large sticks in their elongated bills during migration in case they grow tired of flying and wish to land in the ocean. It was hypothesised that the stick would provide a perching site and allow them to keep afloat atop the rocking waves.

These decorative stories are so intriguing, not only because they map our scientific understanding but also because they show how a species can be so embedded into culture – and in the case of the Woodcock, they're clearly tightly woven throughout our history, as much (or even more so) than they are today.

Have you ever watched a Woodcock in flight? Like clockwork, I'd encounter this sight at 7 pm each summer evening as one flew over my head in the back garden, heading towards the thick woodland. As a crepuscular bird, mostly active at dawn and dusk, they stand out against the setting sun with their erratic, fast flight pattern. It's a special sight to see.

Their arrival back to the UK is known as the 'fall' of the Woodcock, and legend states that the first full moon in November perfectly coincides with their return. Many report their first sightings of the year on this night which is why it has been whimsically named 'Woodcock moon' – a term that echoes down through the centuries.

WORDS – MEGAN McCUBBIN | **ART** – JAMES McCALLUM

Red-necked Phalarope

Growing up in land-locked Shropshire, I could only dream about seeing a Red-necked Phalarope. It was a very rare breeding wader. At the time, in the late 1970s and 80s, there were around 15–40 breeding pairs, virtually restricted to a handful of localities across Shetland and the Outer Hebrides and the occasional nesting in mainland Scotland. It was, of course, a very rare passage migrant to Shropshire too.

My first encounter with a Red-necked Phalarope was on 7 July 1992; a male at Martin Mere in Lancashire. I'd been to Scotland with my friend Chris, and we popped in on the way back to Shropshire. My bird notes remind me it was in "good plumage". I was thrilled to see Red-necked Phalaropes on their breeding grounds for the first time on 25 May 2014, during a family holiday to the Outer Hebrides. I remember the occasion well and my notes tell me "a pair seen briefly before lost in vegetation". The amazing views of a singing Corncrake, displaying Dunlin, Short-eared Owl and Whooper Swan made the experience memorable! We called by later in the day and my notes say "active this evening with five (three females) seen. Two apparent pairs. High, sharp '*tik, tik,tik*' in flight".

One of my most memorable encounters with Red-necked Phalarope was finding a delightful juvenile at Livermere Lake in Suffolk during an evening visit on 21 August 1999 with my partner Pete. The evening light was beautiful, and it was just amazing to watch it spinning around in characteristic fashion. It was my first in the Brecks and a great find locally – only the second record in West Suffolk in the 20th century! We were lucky to find a female at the same site on 30 May 2011.

Red-necked Phalarope remains a rare breeding bird but it's pleasing to see that the Rare Breeding Birds Panel reported record numbers in 2018, with a maximum of 108 pairs across 40 sites. Improvements to the management of their wetland breeding sites may be one reason for the change in fortunes. While Shetland remains the stronghold for breeding, there have been notable increases in Argyll and the Outer Hebrides since 2015, and successful breeding on Fair Isle in recent years. It was exciting to hear about the first-ever breeding attempts in England in 2014 and 2015. With just five modern records, it might be some time before I see one in Shropshire though!

CHANGING BOCC STATUS ❶-❷-❸-❹-❺

RED-LIST CRITERIA APPLICABLE
- IUCN Globally Threatened
- Historical breeding population decline
- Breeding population decline
- Non-breeding population decline
- Breeding range decline
- Non-breeding range decline

WORDS – DAWN BALMER | ART – LISA HOOPER

Kittiwake

It's Bank Holiday at the Newcastle Quayside summer market. I'm knocking shoulders with strangers as kids hassle parents and groups flock for food. Coffee carts, books, vintage jeans ... a vibrant blur of browsing, chatting bodies.

With the river on my left, the shadow of the Tyne Bridge approaches. My eyes follow its iconic curve arching above the pavement, connecting north with south. Moments of white soften the shallow metal edges, and a Mexican wave of tell-tale, namesake calls moves through the heart of the crowd. "*Kitti-waaake, Kitti-waaake*"

My footsteps quicken, eyes to the sky, breath bated, until the coolness of the Tyne Bridge casts over me and finally, I'm transported. Self-awareness drifts as I stand anchored, here in the heart of the city, immersed in a Geordie seabird Mecca.

The Tyne Kittiwakes are the most inland breeding colony of Black-Legged Kittiwakes in the world. This true 'sea-gull' lives a pelagic life out on the open ocean before returning to the Tyne to breed in summer. The spirit of the wild winter ocean flows in their calls, right by Greggs.

A group of lasses heading out pulls my eyes back down to earth as they scream, running under the bridge's danger zone, marked only by the Jackson Pollock style, white-drip pavement that returns by the end of the day.

Others stroll past, at first distracted by phones, chatting with friends. They notice the commotion and pull a Full Emergency Stop. Hoods up, they dare each other to make a run for it under the bridge shouting, "Proper liftin' that!" They reach safety past the bins overflowing by the river. Phew, close call.

I return my eyes skyward, this time on the building by the bridge which netted its ledges; a cosmopolitan cliff to our colonial nesting Tyne Kittiwakes. I can't help but notice a new line of Kittiwakes now nests on the neighbouring building.

Where cultures collide, they can clash. But more often, they mingle to create opportunities for pioneering communities to develop and grow together. The urban coexistence of people and Kittiwakes is no different.

My gaze quickly returns to the curving steel to imagine how we reach this destination, where our seabird metropolis is celebrated year on year. It's precarious, but reachable, and it's ours. I scan the metal ledges; black-tipped wings knock for space as chicks hassle parents, while seafaring fishing groups head east along the Tyne.

WORDS – HEATHER DEVEY | **ART** – ANDREW MACKAY

Herring Gull

My first notable encounter with Herring Gulls was on the Isle of May, an important seabird colony located in the Firth of Forth. I had of course come across them before, their caterwauling so evocative of trips to the seaside. But this was the first time that I'd really looked into their gimlet eyes and felt the intensity of the clouds of gulls screaming for me to get away from their nests. They are fierce defenders of their young and respond en masse to intruders, sometimes swooping so close that you can hear and feel the wind in their wings.

Working on their daintier relative, the cliff-nesting Kittiwake, meant that I often had to pass through the nesting colonies of the larger gulls first, including Herring and Lesser Black-backed Gulls, and their even more imposing relative, the Greater Black-backed Gull. I'd give myself a second to brace myself for the welcoming committee and then step into the fury.

I would see the gull chicks fleeing in panic, diving into the vegetation and taking cover behind rocks. Carefully picking my way though, I then noticed the specked fluffy behinds of small gull chicks poking out from the plentiful Rabbit – and sometimes Puffin – burrows. I would pause, even in the mayhem, to marvel at this comical sight of leopard-skinned balls tucked away in the undergrowth. Clearly they were working on the principle that if they couldn't see me, then I couldn't see them. Meanwhile their parents were alarming from the skies, getting increasingly agitated at my presence.

This is a bird that inspires strong emotions. Synonymous of seascapes, Herring Gulls have become increasingly prevalent within urban areas; here they exploit the food that we waste and nest on the roofs of homes and commercial buildings, rearing their chicks above the bustle of busy streets. Cunning and opportunistic, this is a bird that is a born survivor even when their numbers have been controlled at their traditional nesting sites for decades, largely through years of egg removal due to concerns over impacts on other seabirds. They have found ways to adapt, and now we too must adapt to their presence, learning to co-exist with them in our towns and cities. Do we continue to target their nests through pest control services or are there other ways we can minimise their negative impacts on people, such as better design of refuse bins and the reduction of litter on our streets. Their strong parental behaviour amplifies the interactions with humans but it's a quality I rather admire, even if a little grudgingly at times.

CHANGING BOCC STATUS 1 2 3 4 5
RED-LIST CRITERIA APPLICABLE
- IUCN Globally Threatened
- Historical breeding population decline
- Breeding population decline
- Non-breeding population decline
- Breeding range decline
- Non-breeding range decline

WORDS – LIZ HUMPRHEYS | ART – KATHERINE BOOTH JONES

Roseate Tern

The Roseate Tern is the UK's rarest breeding seabird, and my favourite tern. I first encountered the species during a visit to seabird islands in Maine (USA), when I was visiting to train researchers from the National Audubon Society in the art of deploying miniature GPS loggers on Puffins.

Although those islands hosted numerous auks, the terns were the stars of the show. Their calls filled the sky, and the ground was covered with their nests, each with its speckled eggs, forcing us to constantly tiptoe our way around the islands, while protecting our heads from their relentless attempts to shoo us away. Amongst the vocal, feisty Arctic Terns, and the sturdier Common Terns, the Roseate Terns seemed delicate, almost shy. With the slight pink hue of their plumage, their slender beak, almost fully black, and their long, forked tail, they were very beautiful indeed. But the easiest way to distinguish them from the other terns is by their legs, the manager of the island explained. They have longer legs, which allows them to walk through thicker undergrowth. As a consequence, they prefer to lay their eggs in thick vegetation, or under rocks, which reduces competition for nesting sites with the other terns, which usually nest in more open habitat.

As a scientist specialising in tracking the movements of seabirds at sea, it is always seabirds' extraordinary migrations that impress me the most. And in that field, the Roseate Tern does not disappoint. Thanks to advances in wildlife tracking technology, it is now possible to track the migration of small seabirds like terns by using tiny loggers weighing less than one gram.

A recent study of Roseate Terns breeding in Ireland and the UK used such devices to show that they migrate south all the way to the coast of West Africa and spend the winter in the Gulf of Guinea. Another study, this time of birds breeding in North America, revealed a similarly impressive journey, towards wintering grounds in eastern Brazil. That such long journeys, averaging over 15,000 kilometres return, can be undertaken across open oceans year-in, year-out, by such small, delicate birds, is simply mind-blowing. To put it into perspective: the oldest known Roseate Tern, aged 25 years old, likely travelled over 375,000 kilometers on migration over its lifespan – the equivalent to flying to the moon.

CHANGING BOCC STATUS 1 **2** 3 4 5
RED-LIST CRITERIA APPLICABLE
- IUCN Globally Threatened
- Historical breeding population decline
- **Breeding population decline**
- Non-breeding population decline
- **Breeding range decline**
- Non-breeding range decline

WORDS – ANNETTE FAYET | ART – HOLLY ASTLE

Arctic Skua

As we leave the stillness of Craster Harbour on the incoming tide and make for the Mackerel shoals a mile or so offshore, behind me the smokehouse exhales. The bracing south-westerly, laced with salt air, is chased by the heat of human exhaust; like a living thing driven by instinct, the billowing cloud cascades from the land and joins us in putting to sea. Ahead of us, Gannets fold and plummet from their drifting state, whistling into the grey abyss with poetic elegance and machine-like accuracy, and beyond them, though indistinguishable as they flirtatiously meander, there are definite specks of life, toing and froing across the border of sea and sky, evading my reach and stirring my imagination.

Our boat begins to slow as the engine is silenced. Here the seabed is littered with sunken ships and the surface is scattered with chancing gulls and Fulmars. Together we drift powerless across the wreckage below, the remains themselves a reminder of what the ocean can give and take. Jouncing in the teasing swell, we pitch and roll with each peak and trough, but having not yet developed the legs and stomach required, my eyes remain fixed on the horizon and my hands wrap tightly around the rail until a sudden charge strikes me and my heart begins to race. A lone Puffin propels itself into view. Steadfast and determined it rockets north, homeward bound toward the Farnes, pursued however, by a dark figure. Giving chase, the interceptor darts across the flexing water, rolling right, in an attempt to outmanoeuvre its target. An undefinable rush consumes me as the waves are glanced with the tip of its wing to reveal an unmistakable profile, a pirate's brand. I am gifted just one glimpse, one contrasting flash of white and two protruding tail feathers before the swell and spray mask it's form, and both birds disappear from sight. With us, yet gone, almost within the blink of an eye.

This one vague and unconfirmed sighting could well remain my only, but from it I have taken my own affirmation. In observing the natural world, I witness life at its most pure and with that I have learned to find and understand myself. Out here, I am the understudy. Out here, where sky and sea merge to challenge perspective, where I am helpless and without understanding of location, I am acutely aware of my place. And whether I was graced with its presence or not, I owe a great deal to the Arctic Skua.

I returned that day to the safety of Craster Harbour, with a renewed sense of being.

CHANGING BOCC STATUS 1–2–3–4–5
RED-LIST CRITERIA APPLICABLE
- IUCN Globally Threatened
- Historical breeding population decline
- Breeding population decline
- Non-breeding population decline
- Breeding range decline
- Non-breeding range decline

WORDS – AARON DUFF | ART – TIM WOOTTON

Puffin

'puffin' Puffin' read my lapel badge – a reference to air pollution and climate changes – as I roved the stifling corridors of COP26 in Glasgow. The narrow streets of Hugh Town on the Isles of Scilly, at the other end of the UK, are paved with Puffins. Or at least the gift shops are. This natty, brightly-masked, dumpy, cartoon 'clown' of a bird attracts even the most hard-hearted of holidaymakers, the most iron-souled of politicians.

"Do Puffin beaks fall off?", we ask on Google. No, but like so many of us, they do lose their looks in winter. And their beaks glow under UV light. "Fact!" as my nephew would triumphantly say. On the Puffin's narrow shoulders rests much of the burden for the UK's love affair with its seabirds. Me, I prefer the edgier Gannet. A pencil case with a Gannet on would take your fingers off with its zip. But, a bit like the Beatles, everyone comes back to Puffins in the end. Go on a seabird cruise. Everyone stands up for the Puffins. Even their scientific name means 'little brother of the North', in their black and white friar's robes, feet held in prayer as they take off.

But beyond the surface charisma, the Puffin is the end of pier comedian with a startling hinterland. Diving up to 60 metres deep, they swim as though they are flying. They dig their burrows using first their beaks but then their legs, like a dog. They mate for life and they supply their pufflings (I know!) with fish up to 24 times a day, using an ingenious set of denticles to fill their beaks to bursting point. The RSPB ran the 'puffarazzi' citizen science project to gather photos of Puffins with filled beaks over time. Why? Because we think changing sea temperatures and pressure from fisheries is leading to a shortage of food. And birds that flap their wings up to 400 times a minute need a lot of sand-eels, Sprats and rocklings.

Like so many of our seabirds, this popular frontman is hovering on the brink. If the food shortages don't get them, pollution events and ground predators (Rats, Mink, Cats) will. If we want our Puffins to be more than jolly pencil case illustrations, then sustainable fishing, protection of feeding grounds, considerate placing of offshore wind farms, a reduction in marine pollution and preventing ground predators from reaching nesting colonies are what's needed. "Here all summer" says the end-of-pier comedian. Not if we don't sort it out.

WORDS – BECCY SPEIGHT | **ART** – DAVID HALL

Hen Harrier

Waiting for Margot.

The curtain rises on a windswept upland. There is a single tree. Offstage a voice reads out a list of names ... *Saorsa, Tarras, Marc* ...

"How long have we been here?" "**Here, in this same place?**" "In this same place." "I can't remember. Sometimes it seems it might just be today and sometimes it seems it might be years." "But in the same place?" "**I would say so.**" "Not moving forward, maybe moving back, but not moving forward?" "**No, not moving forward.**" "Waiting for Margot." "**For Margot and all the others.**" "I don't see them coming back ..." ... *Rannoch, River, Blue* ...

"What do you make of that sign?" "**The one spinning from the hollow tree?**" "The one that says 'Nothing to see here'. Does it mean there is nothing to see here because nothing can be seen ... ?" "**Or that there is nothing to see here because there is nothing left to be seen?**" "Perhaps both." "I don't suppose we're meant to know for sure. That's the way of words up here." ... *Vulcan, Skylar, Rain* ...

"The wind is harsh out here on the moor." "**It is. Cold, angry and unyielding.**" "It's not natural." "**For the birds?**" "Yes, for the birds. What else matters?" "**They seem like candles born into the teeth of a storm. The sky dances above their wings for just an instant before the raging gale snuffs them out again.**" "Even the ghosts have left this place now." ... *Lia, Hilma, Yarrow* ...

"Do you see that figure?" "**The one half-hidden but watchful? Keeping in the shadows?**" "Is that an arm raised in triumph and pointing to the sky?" "**Burnished. Ramrod straight ...?**" "Or something else?" "**Nothing is clear up here. Everything shifts and curls like smoke.**" "Deliberately so?" "**Part of the game.**" "Ah, yes, it's all about the game." ... *Fingal, Silver, Solo* ...

"I can't go on like this. Sitting and talking. We should move on." "**But if we go who will remember Margot, the others, all of them. Who will applaud their lives? Who will even know they were here once, if not us?**" "And they were here once." "**They were here. And should be here now. If we leave the stage the curtain could fall and never rise again.**" "Then we mustn't. But we can't just sit and talk." "**No. That's not enough. Not any more.**" "A storm of our own?" "**A storm of our own.**" ... *Margot* ...

WORDS – CHARLIE MOORES | **ART** – PHILIP HARRIS

Montagu's Harrier

I've been obsessed with birds of prey ever since I saw my first Red Kite in the Elan Valley in the 1990s. Wherever I go, my eyes and ears are constantly scanning the skies, fields and trees for owls, kites, hawks, harriers, falcons and now Ospreys and eagles thanks to reintroduction projects in southern England. It feels like a dream that I get to work closely with these majestic birds.

Every breeding season has its highs and lows but 2014 stands out as one of my favourites. Owing to a glut of Field Voles, it was the best Barn Owl breeding season on record. And it was during that summer's fieldwork that I had my first encounter with the UK's rarest breeding bird of prey.

On 28 August 2014, as I perched on the old wooden beams of a stone barn in Wiltshire, an elegant long-winged grey and black plumaged bird flew through my peripheral vision. I carefully returned three downy owlets to their nest box then hurried down the ladder to reach for my binoculars. With the help of agitated Buzzards and corvids, I spotted my first ever male Montagu's Harrier sitting atop a neat tower of straw bales. And he looked breathtakingly beautiful in the golden late August sun.

My heart raced with excitement as I watched him inspect every inch of the surrounding freshly cut stubble field with his vivid yellow eyes. Unphased by mobbing birds, he stayed there for an hour, as did I, only leaving once our most graceful harrier was no longer visible. You see, I wasn't sure when I'd next see a Montagu's Harrier and I haven't seen one since. That was eight years ago.

Within days he would have left our isles to embark on his autumn migration back to sub-Saharan Africa for winter. Nine pairs bred in the UK that year. Five of these were in Wiltshire, two in Norfolk and one each in Lincolnshire and Yorkshire. I wondered if my male had bred that spring.

Sadly, the Montagu's Harrier is a new addition to the Red List. Its UK breeding numbers have been precariously low for a number of years, and despite targeted conservation work, the number of breeding pairs has rapidly declined since 2015. Devastatingly, there were no Montagu's Harrier nest records in 2020 or 2021, with many fearing that it could soon become extinct as a breeding species here in the UK.

CHANGING BOCC STATUS 0-2-3-4-5
RED-LIST CRITERIA APPLICABLE
- IUCN Globally Threatened
- Historical breeding population decline
- Breeding population decline
- Non-breeding population decline
- Breeding range decline
- Non-breeding range decline

WORDS – EMILY JOÁCHIM | ART – JON TREMAINE

Lesser Spotted Woodpecker

Over the last few decades this beautiful little woodpecker has quietly slipped away from our regular birding experience. Many birdwatchers have given up on seeing them altogether, or must make a pilgrimage to places like the Brecks or New Forest just for a glimpse and a snatch of drumming.

It was not always like this. In Victorian times it was thought to be the most numerous woodpecker in many areas. As recently as the 1970s, during the Dutch Elm Disease episode, the Lesser Spotted Woodpecker could be seen and heard widely; the sight and sound of them chasing each other through the canopy, calling and doing their butterfly display flights, would have been a common experience. Seeing something like this nowadays makes for a very special birding day. Even in the 1980s when we first started our woodpecker studies there were Lesser Spots breeding in all four of our Hertfordshire study woods. But not now – they may just be hanging on in one of the woods.

RSPB research in the early 2000s implicated low breeding success as the driver of Lesser Spot decline but even with professional full time fieldwork it proved extremely challenging to collect enough data. In 2015 we created Woodpecker Network realising that, with a little help, volunteer birdwatchers could make a big contribution to our knowledge and conservation of this species. Since then, more Lesser Spots have been reported, and the numbers of nests monitored and reported to the BTO's Nest Record Scheme has grown to 10–20 each year … and, most importantly, we have attracted and helped a group of dedicated Lesser Spot recorders.

Our monitoring has confirmed low breeding success and that this has declined over time. We have also demonstrated clearly that low chick survival, especially when the young are very small, is the big problem and this is probably related to the habitat, food and feeding. We are sure the Lesser Spot is still in deep trouble but with an estimated 2,000 pairs during the 2007–11 atlas there are probably more of them around than was thought. They are so unobtrusive that without dedicated observers they are easily overlooked.

Our increased understanding of the factors driving the decline of the Lesser Spotted Woodpecker is just the start. The challenge now is to turn this knowledge into practical conservation measures on a large enough scale to turn the situation around. We are making a start but there is still much more to do.

CHANGING BOCC STATUS 1 - 2 - 3 - 4 - **5**

RED-LIST CRITERIA APPLICABLE
- IUCN Globally Threatened
- Historical breeding population decline
- **Breeding population decline**
- Non-breeding population decline
- Breeding range decline
- Non-breeding range decline

WORDS – KEN AND LINDA SMITH | **ART** – MACKENZIE CROOK

Merlin

I was on my friend's stag weekend in Wales when I saw my first. My decision to bring binoculars hadn't proved popular but I felt vindicated when I pointed out the silvery blue bird hovering above Ysgyryd Fawr. "Merlin", I murmured, and the young men nodded appreciatively before handing round a hip flask.

My dad is a proper birder. He's who I was emulating with my binoculars and he's seen many a Merlin over the years. When I asked him what the bird conjured up in his more mature mind, he instantly reeled off the key facts:

"Our smallest falcon, male smaller than the female, not much bigger than a Blackbird. They're birds of the northern moors, breeding there but coming further south for the milder winter weather down here. In falconry, known as a 'Lady's Hawk', popular with Mary Queen of Scots, Catherine the Great, et al. Little males particularly smart, hence prized as a bit of a fashion accessory in the bad old days. Problems they've faced: loss of habitat, pesticides and persecution on the grouse moors, numbers have improved but still rare and one has to be really lucky to see one, either flying or perched. I've seen single birds on the Downs (chasing a Skylark), zooming over Pulborough Brooks, perched near the beach at Climping, and on open ground on the Isle of Sheppey – all typical places in winter. Can turn up anywhere in open country, and only usually seen momentarily as they dash around extremely quickly, chasing smaller birds like Skylarks and pipits."

"That's what I thought", I said. "Such magical birds of prey. I'll never forget my first." I recounted my Welsh find, the hairs on my neck prickling at the memory.

"They don't hover", said Dad.

"This one did", I said. "They categorically don't hover", he insisted.

It turns out I've never seen a Merlin. I saw a Kestrel on that stag. And so the Merlin remains on my long list of never spotted species, alongside Capercaillie, Nightingale and Golden Eagles.

But I still treasure my non-Merlin. I got so excited about the sighting; it genuinely made the stag weekend for me and my mates. And what difference does it really make if we saw one or not? I was sure I'd seen a Merlin, and maybe that's enough. It certainly makes the hobby more fun if you can be a little relaxed with the truth. So here's to the boasters, the believers, the bad birdwatchers. And speaking of hobbies, I think that's one on the bird feeder right now …

WORDS – ALEX HORNE | **ART** – RUTH WEAVER

Red-backed Shrike

The first time we saw our home, Wild Finca, a dilapidated old cattle farm in Asturias in north-west Spain, Early Purple Orchids had peppered the verge alongside the entrance track. And as we'd walked the land a huge shadow had blocked the sun, a Griffon Vulture soaring low overhead. It didn't matter to us that none of the buildings were viable for human habitation, it was its wildlife potential that had us spellbound.

Roll on spring 2019, and the previously heavily grazed fields had begun to recuperate, and with the habitat came the migrants. First the Grasshopper Warblers turned up, reeling away from the dry Fennel stems amongst the Gorse. Then the Egyptian Vultures appeared, patrolling the hillsides, while the Swallows swept into the barns to rebuild the previous year's dwellings. Gaudy Common Redstarts, fresh from crossing the Sahara, alighted on our old Walnut tree to serenade the sunrise. Secretive Golden Orioles piped away from the canopy, in a constant game of hide and seek. And then the bandits arrived.

It took us a moment to work out who the latest arrivals were. From the house we could hear the harsh call, repeatedly shouted from the tops of the highest trees around the land. A binocular scan from the living room window finally revealed him, but not for long. Jet black bandit mask, granite grey helmet, thick hooked bill, brilliant white breast, and he had turned to reveal that quintessential sleek conker brown cape.

The male Red-backed Shrike swooped down into the meadow below, disappearing behind a tuft of grass, before reappearing, his bouncing flight taking him to a nearby fence post. A flick of the head and a vibrant Great Green Bush-cricket disappeared down his throat. Another swoop down, this time returning with a shiny black dung beetle. A truly mesmerising bird.

Once we'd spotted one, we saw another, and then another. From the fence lines, to the bramble pillows. Even the veg patch became a hunting ground for the Red-backed Shrikes, our bean poles newly adopted vantage points for their hunting forays.

We built a map of their territories and once the chicks fledged it was impossible not to see them, the rowdy offspring chasing their parents relentlessly, a riot of chattering. Followed by silence. Back south they headed, and until the following spring we are always left waiting, wondering, will they come back?

WORDS – KATIE STACEY | **ART** – JANE TAVENER

Marsh Tit

If I stretch in my seat, I can just about see over the pathway's scrubby growth to the pristine, distant marshes; its mosaic glints of water, the temptation of an open sky. It pulls at my chest because it's forbidden to me, that's the truth of it, but believe me: I've still tried. My wheelchair carries old mud from our laughing, failed attempts to navigate the trails that lead to it, sinking, tipping, retreating, as birders pass with long strides and longer lenses, talking of Lapwing, Little Egret, Cormorant.

I must stay away, here in the dry by the litter and the trees. I tell myself that's no bad thing. Great Beeches grow old and stubborn along the edge of the industrial land I'm stuck in, throwing their high, grey arms out over the wild, untidy hedgerows that persist in thick strips and scraps between things, overlooking gardens, warehouses, barbed wire. There are hard paths with hidden entrances that are good for wheels: a maze of urban nature with its own rewards, for it is in this jumble of an edge place, edge-being I am too, that Marsh Tit finds me.

"*Pit-CHOO!*" It sidles up and sneezes by my side, warm and sensible in its woolly black cap, afraid of catching cold.

"Hello again" I say as we both fidget, awkward, because I know its secrets and it knows mine. I know its name and I don't think it's used to that. I know too that despite that name, it hates the damp and loves the safety of the old, dirty trees; in turn, Marsh Tit seems to know that despite my pragmatism, I am lonely, that I long for the sky and for open space. It knows I'm not really satisfied with any of this.

Does it know it is a rare bird? I have told it so quietly sometimes, not wanting to cause embarrassment. I have told it many other things, too. I have said, throat tight, that it means the world that it comes to visit me, here, right here where I am. Me, who feels so often left out of the conversation about this bird or that bird because I can't get to see the bloody things, and yet I still love them with all of my heart.

"*Pit-CHOO!*" it sneezes again, sensitive, small and overlooked, like me.

"Bless you," I say, and mean it.

WORDS – JOSIE GEORGE | **ART** – HELEN MUSSELWHITE

Willow Tit

Never mind its dull cap, its pale secondary edges, its on-average-larger bib. It's a bull-necked, plain-clothed, rough-voiced hewer of wood; a no-bullshit plain-speaking rough-diamond worker. Keeps itself to itself. Doesn't prance about for all to see (unlike some). Gets on with stuff, checking for rot, hacking out test holes, that sort of thing.

Hides in plain sight; went undercover for 221 years, from the *Ornithologiae* of Francis Willughby until outed by Ernst Hartert and Otto Kleinschmidt in 1897. Subverted the old order: field operatives scattering to re-check old records of Marsh. Still keeps a low profile. Frequently goes missing.

Here in Yorkshire they were "partial to the combination of woodlands and water, particularly in river valleys and places where brushwood is found, and low-growing trees afford suitable sites for the nest", as Thomas Henry Hudson wrote in 1907. No matter that he thought he was referring to Marsh Tits. So did Seth Lister Mosely of Huddersfield who, in 1915 wrote that he had seen it "in all our moist woods". By 1936 careful observers like Tom Johnston in Cumbria were starting to notice the differences in habit between the two species: "the preference shown by the Willow Tit for damp and even swampy ground is explained by the fact that the trees, especially birch, growing in such situations, are most affected by decay."

Ninety-four per cent of Willow Tits have gone missing since then. Alex Lewis compared woods that were still occupied by Willow Tits with those they had abandoned. She found one difference: the soil was wetter at occupied sites. The British countryside, including its woodland, is drying out.

Here, we seek it in the damp recesses of industry, the wet spoil slacks, the subsidence flashes. Look in the sallow-birch-tarmac community, the ecological ersatz. Look to places where the rot thrives, where a soft stump can be hollowed and a hard living made in the tangle of bramble and twisted angle iron. For while no bird ever evolved to live in closed-down collieries, it is here that a motley assemblage of refugees from the failed states of nature gather.

If the Willow Tit would listen, what counsel would I give? Go west. Go high; I am told the woods are wetter there. You were not made for the human world, and they say the second realm of the Beaver is at hand. Try to hang on a few more years.

WORDS – LAURENCE ROSE | **ART** – FAITH CHEVANNES

Skylark

From high above the cornfield
Come beeps and cheeps and trills
That sound to me like Techno
From DJs with computer skills

But it's just a small brown bird
Riding up and down on air
They call this guy the Skylark
And as names go I think that's fair

It ***does*** look like they're having a lark
And there's no doubt they're in the sky
But I'll always call them 'Techno Bird'
We've already established why

In case you don't believe me
Check out Higher State Of Consciousness
Towards the end you'll hear the sound
Of a Skylark, more or less

(Though it's actually a Roland TB-303 bass synthesiser)

CHANGING BOCC STATUS 1–2–3–4–5

RED-LIST CRITERIA APPLICABLE

- IUCN Globally Threatened
- Historical breeding population decline
- Breeding population decline
- Non-breeding population decline
- Breeding range decline
- Non-breeding range decline

WORDS – ADAM BUXTON | **ART –** ADAM BUXTON

Marsh Warbler

Knowing that a Marsh Warbler had been reported in a hedgerow on an autumn twitch made it no easier for me to be sure that I was not looking at its common cousin; skulking behaviour hid diagnostic features – Marsh or a Reed? Simple recognition would be a joy.

Some birds are special because of their colours; flashy Rainbow Lorikeets or pastel-shaded Abyssinian Rollers. Some are attractive through their displays; we thrill to the strut of a blackcock lek or the sudden fanning of a Peacock tail. Some birds charm is their cryptic plumage – the unexpected discovery of a Nightjar in leaflitter or the exquisite patterns of a closely observed Wren. Some bring tears from the years of wishes fulfilled, as when a Wallcreeper slapped into a rock, feet from me as I crossed a shallow stream-bed in India; a dream coveted since the day I collected its portrait in a tea packet nearly half a century before.

Another 'cigarette' card collection called 'Curious Beaks' made me wish for hornbills, Shoebills and toucans; who can deny that they are won over by an Avocet's upturned bill, or the sideways tweak of a Wrybill's.

Bellbird calls echoing through rainforest enthral. Others stun you with their vocal range and repertoire ... I stayed awake for hours when woken by a Nightingale serenading a potential mate.

Knowing that rarity makes for limited viewing opportunities, evokes joy when you see a vagrant American wader in the UK or an albatross where it just should not be.

My joy of recognition was fulfilled years later. My son pointed low down, almost at the foot of a sturdy weed. There a bird busied itself building with inbred skill. His young ear could separate its call from the monotony of ubiquitous Reed Warblers all around Two decades and half a hemisphere away, a guide had me listen to an African Marsh Warbler, then to its visiting relative. Even my tin ear could hear they were different.

Recently, my son sent me a video clip. On his inland patch someone had found a Marsh Warbler male and my son found its mate and recorded their courtship. Visually, I could barely see how experts tell them from their abundant cousins and my ageing ears no longer tell Rooks from Redstarts. However, I could bathe in the reflected glory; recognising his skill and sharing his joy.

WORDS – BO BEOLENS | **ART** – MARK ANDREWS

Savi's Warbler

From the heart of the reedbed came a long, rolling ticking sound, like a fingernail on a comb. Rising and falling in volume.

Was it a Grasshopper Warbler (or 'gropper' in birder speak)? We record groppers on the reserve most years, but this sounded harder and louder, easier to place.

Yes … there! Head-height, near the top of a stem, smack in the middle of the densest bit of reed monoculture, with its beak fixed wide open and that strange, dry sound streaming out constantly; the volume adjusting with every turn of its head.

Such excitement – it was definitely too plain for a gropper. In fact could there be a plainer looking bird? Probably the most distinctive thing about its appearance is the fact that there is absolutely nothing distinctive about its appearance. Savi's must be the ultimate LBJ or 'little brown job' – wow, what a beauty!

Word got out and the county bird recorder was there within half an hour to breathlessly confirm that it was a 'county first' for Carmarthenshire, and he was followed by a steady stream of twitchers. One man was so impatient that he queue-barged through reception without paying the entry fee, waving his tripod angrily. The centre manager threatened to call the police but the man blustered on, leading to the vision of eight police in full riot gear walking quietly through the reedbed, and the stunned man never even saw the bird.

We heard its amazing song on and off for two more weeks in the bright early mornings and it created a real atmosphere in the rustling, moving reeds. Being so far from its normal breeding range (East Anglia is considered the westernmost limit), this male's song carried a sense of melancholy, with the thought that he was so unlikely to find a mate all the way over here in windy south-west Wales. But it also carried hope – the hope that this bird might be a pioneer, a lonely foreteller of a possible westward spread in the distribution of Savi's Warblers.

WORDS – BRIAN BRIGGS | **ART** – JEFF BAKER

Grasshopper Warbler

My first binoculars were too heavy hold to my eyes for more than 10 seconds. But that was enough to see a Blackbird sing. I'd heard Blackbirds, I had seen them, but never put sight and sound together in such detail. I watched his throat swell and his beak open to let out the sound. Vision, hearing, emotion lost their boundaries in that moment; I felt the notes, I saw them in the air, like raindrops defying gravity, moving upward into the spring sky.

After that, seeing a bird sing became a small, intense addiction, delivering a hit of deep cross-sensory joy. If I heard a bird singing, I had to get close enough to see it. There were plenty of easy fixes to be had; Wrens almost breaking with the volume of their song; Robins, closed beaked, as if singing to themselves; Willow Warblers, tumbling sound from the tallest trees. But seeing the singers who don't like spectators provided a special sort of trip, one I was prepared to work for: two hours spent peering into a bramble thicket to disentangled the silhouette of a singing Nightingale; an afternoon circumnavigating a willow copse, where a Cetti's Warbler taunted me invisibly; long, Hebridean dusks listening at every field edge, finally rewarded with a Corncrake's small, red mouth opening to say its scientific name.

There was one shy singer that always eluded me. I heard my first Grasshopper Warbler on the Brecks of the Norfolk/Suffolk border when I was 17. It was a liminal sound, like something half remembered from a dream; so insect-like and even for young ears, already on some sensory edge. Back then I didn't see the bird, so I decided I was mistaken.

In the almost half century that's passed since then I have learned to trust my senses and to be grateful for whatever they deliver. When I heard the soft, susurrating churr – the very embodiment of incessant – on the abandoned airfield near my home, I knew it was a Grasshopper Warbler. It was dusk, the vegetation a smudge of sage and indigo, the sky rosy as a child's kiss. The whole place seemed to listen, and I stood still and listened too. I didn't try to get close, the light was fading in any case and the moment so fragile I didn't want to break it. It was enough to hear this rare, strange sound filtering through from some other world of avian consciousness.

Grasshopper Warbler song is at the frequency that human ears lose first as they age. Soon, its sound will leave me, just as the bush-crickets have already left. Will you listen to it for me? Will you make sure it stays in this world and doesn't slip forever into another?

WORDS – NICOLA DAVIES | **ART** – KATHRYN O'KELL

House Martin

Despite the unpromising aesthetics of the dense tower blocks and estates, located here on the south side of the Thames in east London, the House Martins are back. Arriving in April, the constant coalescing and dispersing of small social groups of feeding and courting birds, and semi-colonial living, closely matches human activities; the birds' noisy chattering a perfect accompaniment to the shrill cries of playing children, neighbours squabbling and diverse music genres issuing from open flat windows. Do House Martins in quiet neighbourhoods of middle England behave more sedately, I wonder?

Although they don't often get a close look, people here do know these birds from their familiar calls and flashes of black and white plumage as they perform aerial acrobatics overhead. Some know more – one long-term resident of the block raises the blue with red-striped flag of the Democratic Republic of the Congo from his balcony every spring, on the week the House Martins arrive, in recognition of their time in Africa. He may be right about the Congo, but we actually know little about their exact wintering quarters. Assumed to migrate to somewhere in sub-Saharan Africa, House Martins are too small and too dependent on their aerial dexterity to safely carry the types of tracking devices currently available.

UK House Martins are largely, but not completely, urban birds, building short funnel-shaped nests of mud or occupying artificial nests under the eaves of buildings. They are born procrastinators; arriving as early as March but then spending weeks apparently socialising, before finally starting to breed a month or so later. In autumn they are similarly reluctant to leave on their journey south, often trying for one more late summer brood and hanging around until the end of September.

Bird conservation organisations have been concerned about the status of this species for years, but steep declines of more than 50% in numbers on BTO/JNCC/RSPB Breeding Bird Survey sites over the last 25 years have put it unequivocally on the Red List. A similar pattern in other aerial insectivores, such as Swift, implies that reductions in the numbers of flying insects is a key factor, though changes in food or habitat over the winter or during migration likely play a role. Modern building construction and the switch from wood to plastic facings may have reduced nesting opportunities, but other buildings have been improved by providing artificial nests. We must do what we can to help these most delightful of urban neighbours, so that our urban landscapes continue to bustle with their social interactions.

WORDS – DAVID NOBLE | **ART** – JAMES GREEN

Wood Warbler

In the long winter months of January and February, the sound of our summer migrant birds seems an unbelievably long time ago and the return of our sub-Saharan visitors an eternity away. On these dark, cold nights, I have the ultimate antidote and the perfect pick me up as a way of letting some light and warmth back – my sound library. When going through my archive of birds I invariably find the cursor of my laptop way down at the bottom of my list, hovering over the song of the Wood Warbler.

It was an early May morning in mid-Wales. I was on a west-facing slope in a forest of wind-blown oaks, halfway between Aberystwyth and Machynlleth, when I recorded my first Wood Warbler. The weather was good, and I stopped in a sunlit glade with an open canopy and very little ground cover, perfect habitat for this little bird. As always with me I heard the bird before I saw it, which is exactly the purpose of the male song – to attract a mate and stake a claim to his bit of forest. He was high up, flitting between a couple of song perches, and as I looked up I could make out his distinctive white breast and lemon coloured throat; a lovely looking bird with an amazing song.

The best way to describe the song is top think of a spinning coin on a table top; as the coin stops spinning so the call fades away in a descending tone. What is also extraordinary is the amount of effort put in to producing the song, his body replicating the blurred motion of the spinning coin! His lower mandible becomes impossible to get a fix on as it moves so fast as the powerful trill is emitted through the mesmerising motion, which then moves to his tail and wing tips adding to the visual effect.

Having got my recordings, I slumped to a base of a tree and just listened, for how long I don't remember, but when I play the recording now it takes me straight back to that spot on the west-facing slope.

So now as I sit here looking out of my window on a grey January morning with my laptop in front of me, I press the space bar again and again, playing the recording with hope that our Wood Warblers will return again soon.

WORDS – GARY MOORE | **ART** – MILES CLUFF

Starling

As I sit here, on a bench, looking out to sea, I ponder one question. How is it possible that Starling populations are vulnerable?

In Shakespeare's *Henry IV, Part One*, (Act 1, scene 3) the headstrong Hotspur, enraged that the king has called his brother-in-law, Mortimer, a traitor, declares: "I'll have a Starling shall be taught to speak / Nothing but Mortimer and give it to him / to keep his anger still in motion." He is fantasising about teaching a Starling to repeat Mortimer's name, to drive the King mad.

There are three interesting things about this. The first is that a Starling could be a gift for a king. The second is that Starlings can be taught to speak. This was common knowledge 400 years ago, which is why Shakespeare uses it as a device in a play. Starlings have interesting songs because they will mimic other species. In May 1784, Mozart purchased a pet Starling and taught it to sing part of one of his piano concertos. He loved the bird. It died a week after his own father's funeral, which he did not attend. He gave the bird a full memorial service, a tombstone and wrote it a poem.

The third interesting thing about this one single mention of the Starling in the Shakespeare canon, was that it is responsible for all the European Starlings in America. In 1890, an eccentric German chap called Eugene Schieffelin, a member of the American Acclimatization Society which introduced plants and birds from Europe to make the new nation of America feel more like home, took it upon himself to bring the 60 species of birds in Shakespeare's plays to the United States. On a snowy, cold spring morning in March he released 60 Starlings in the middle of New York City. Now, the US population of Starlings is estimated to be 200 million.

The Starling has a powerful mystique, exuding a wild intelligence. Were I an evil villain, with a dastardly plot, I would want my attire to ape their plumage. I would wear a cloak of dark feathers shimmering with iridescent purples and greens; a hypnotic firmament tipping the quills.

When the world feels grey or mundane, I go to the West Pier in Brighton, a collapsing pick-up-sticks architectural construction being devoured by brine. It is the roost for a grand choir of thousands of chattering Starlings. On some invisible signal, they will launch into the sky, performing an incomprehensible synchronised dance, designed to impress, overwhelm, and confuse predators. It never fails to ignite wonder in the watcher. The undulating cloud of birds throws shapes, a double helix, a giant albatross, a jellyfish, each Starling but a spec. The noise is as crashingly cacophonous as the angry sea. This is the creature that gave us the murmuration.

What monsters must we be, that we have reduced it to sit on the Red List.

WORDS – M. G. LEONARD | **ART** – BEN ROTHERY

Mistle Thrush

I was about 12 when someone stuck their head into my year seven history lesson (Suffragettes, in case you were wondering) and asked 'Does anyone know anything about birds?' It was as if Peter Parker had been asked if he knew anyone who could walk up walls. Yes! At last! My obsessive reading and re-reading of *A Field Guide to Birds of Britain and Europe* **had finally paid off! I stuck my hand up.**

It turned out that a bird had flown into the window of the dining-hall, leaving one of those ghostly imprints on the glass. It was lying on the gravel outside and, although someone had found a cardboard box, no-one wanted to pick the bird up. I did. It was still alive, panting and dazed, larger than a Blackbird, pale-headed, its chest and neck thickly and randomly freckled: a Mistle Thrush. And how perfect it was, close-up. My *Birds of Britain and Europe* might fall open at the page entitled 'Rarities'; I might live in hope of seeing a jewelled Bee-eater in my Midlands back garden, but this modest-looking bird, all shades of brown, almost weightless in my hands, was as beautiful as any exotic blow-in.

I was allowed to take it home. I walked back carefully, dreaming of a tame pet that would come when I called, but it woke up at the bottom of my road and started hurling itself around the box. I opened the flap and it was gone, like a subtly-coloured firework.

I liked to think that it was grateful, though, and that it hung around afterwards. Do you see that bird tearing into the Holly berries and defending its favourite bush against all-comers, or swaying in the cold wind at the top of a tree while singing as sweetly and mournfully as any Blackbird? Do you see that bird poking around the dead leaves in the park, seemingly greyish-brown until it takes off and flashes its white armpits? Do you hear that bird calling with a lovely soft rattle that sounds like a pencil pulled across a comb? That's my Mistle Thrush, that is …

WORDS – LISSA EVANS | ART – HAYLEY CHAN

Fieldfare

"That's not a thrush …" I whisper the words out aloud, though no-one is here to listen. It is early February and our garden is all but white; earth barely distinguishable from sky, and the view from the window clouded by snow which has been falling steadily since first light.

The bird who is "not a thrush" is tugging at a partially buried apple which, last time I looked, Blackbirds had been squabbling over. Each year we keep back a few tubs of apples to feed the birds over the cold winter months – and each year the Blackbirds spend far more time guarding them and chasing each other away, than they ever do eating them.

New to birding, it takes me a moment to be sure I have correctly identified our visitor. A quick scan of the garden reveals he is not alone; in fact three more of his kind are in the Hawthorn. I have seen Fieldfares before in fields and tree tops, usually amongst flocks of other winter thrushes, but they tend only to visit gardens in prolonged inclement weather, so this is indeed a treat. I pick up my binoculars and hone in on the bird on the ground. My goodness he's handsome! Slightly larger than the now displaced and rather disgruntled Blackbirds, his breast is speckled like other thrushes, but he has a blue-grey head, chestnut back and wings, snowy-white underbody, and a saffron-blushed throat; an altogether 'cooler' palette than his relatives.

These beautiful winter nomads travel from breeding grounds that extend across northern Europe, including Scandinavia (where they are known as 'birch thrushes', or 'björktrast'), to take advantage of our milder winters, and to feed upon our berries after their own have run out. Arriving on our eastern shores from mid-September onwards, they roam around the countryside in search of food, and living up to the name 'Fieldfare'; likely from Old English 'feld' (field) and 'fara' (to go), and interpreted as the 'traveller of the fields'. Whilst favouring a diet of berries from trees such as Rowan, Hawthorn and Holly, you are just as likely to find Fieldfares in open farmland, where they feed on earthworms and other invertebrates.

Prompted to find out why Fieldfares are on the Red List, I discover that milder temperatures in northern Europe, combined with unfavourable winds for migration, may have led to fewer making the journey to overwinter in the UK. Knowing this makes me cherish today's sightings all the more.

WORDS – BRIGIT STRAWBRIDGE | **ART** – CHARLOTTE STRAWBRIDGE

Ring Ouzel

We tracked her from her wintering grounds in the Moroccan High Atlas and spent tense days as she weathered Murcia, Valencia, Aragon, sticking to the plateaus and flying mostly at dusk. When the GPS blipped out near Pau we tried not to imagine her clamped in the grin of a gundog or strung around the waist of some rifle-toting bâtard in fatigues. It was fist-bumps and champagne-popping emojis when she came back on the screen, and we swore next year we'd pay the extra for an upgrade to satellite tags. Next was a risky short-cut over Biscay but from then she made great time, hitting Hampshire, we reckoned, on the last day of March.

We relaxed a bit then, and had time to reflect. You said that even in our grandparents' day the birds might have stayed in southern England, to be seen by workers or wanderers in the Purbecks or on the Dorset Downs, who perhaps paused to look twice at a Blackbird so high up. That's when they'd have seen that tell-tale half-moon torque, bright white against the black plumage of the cock, looking more like a rime of limescale on the hen. Nowadays though, they're retreating to the north and west, and are rare enough even there.

What you said next gave me goosebumps: that all over the world glaciers, too, are retreating up the mountains, and that the day will come when we'll do the same. We'll leave the valleys and again confine ourselves to hillforts, fleeing our own barbarity, fearing our own shadows. Then you broke off for a long, dramatic drag on your overworked vape.

A week later our bird was in the Berwyns. But by then we wanted to see one for real. You had to work on the chosen day, but you gave me your envious blessing for the bus ride through the Beacons to Crai. I skirted the reservoir and climbed through skeins of birdsong to the shining moorland of Waun Leuci, where Skylarks ran through the twisted grass and a pair of Ravens honked me off their nest in a pine by the pillow mounds. I was impatient, overkeen, scanning every outcrop, every wall. Eventually the beauty of the place slowed me down and then halted me altogether. Empty of thought I watched the sun drag a golden cloth across the four hundred-million-year-old slab of Fan Gyhirych. It was so quiet, as though the whole landscape was waiting for one absent.

CHANGING BOCC STATUS 1-2-3-4-5
RED-LIST CRITERIA APPLICABLE
- IUCN Globally Threatened
- Historical breeding population decline
- Breeding population decline
- Non-breeding population decline
- Breeding range decline
- Non-breeding range decline

WORDS – STEVEN LOVATT | **ART** – STEWART SEXTON

Spotted Flycatcher

Behind the tea-towel-cosy image of your typical National Trust cafe are the teenage catering assistants, breaking a sweat to meet the pressures of providing scones en masse to the coach party hoards; at which point only the strongest survive. My 19-year-old self learned this the hard way during a season at Mottisfont Abbey in Hampshire, but I knew that salvation lay in the lunch break. Not just for the obvious reasons, but because it was my time to catch up with a delightful pair of Spotted Flycatchers who had chosen the old wall beside the staff room window as their nest site.

"Zip-flip-zoop. Zip-flip-zoop". That unmistakeable ball-on-a-string tactic of, erm, flycatching was, like nature is for so many people, a grounding joy that shined bright from within the daily grind. And people have the nerve to call this bird a 'little brown job'! For one its plumage is silver, turning to flecks on white, in what I would term waterfall chic. A spritely character, as enthused and enigmatic as Puck from *A Midsummer Night's Dream*. And when you factor in their epic migration to the rainforests of Africa, then 'little brown job' is nothing less than offensive.

Unfortunately, like so many of our birds, its mystique is no doubt imbued with a trait that no one wants to add: rarity. It is no longer sprite-like simply because of its behaviour; now, it is also sprite-like because it has become a last vestige of magic, occasionally appearing in a world where that's rapidly fading.

Whenever this bird shows up in my life it's joyful, but I'm always left wondering when the next time will be. The last occasion was in 2019, when a pair set up shop in the woods at which I was working as a ranger, woodlands within which Bristol's Bear Wood exhibit had been built to recreate Britain's lost woodland landscape. I could witness the flycatchers' *"zip-flip-zoop"* at eye level from the walkway that wound its way through the canopy, the birds nonchalant about the bears and wolves that roamed below within their enclosures. I shouldn't have been surprised; after all, they have gorillas and elephants for neighbours in winter.

Three years on, I'm still waiting for my next meeting. The optimist in me hopes they will become more frequent. But with the way the world has been turning, I'm also regretfully admitting that the reverse is equally likely and probably more so.

CHANGING BOCC STATUS 1-2-3-4-5
RED-LIST CRITERIA APPLICABLE
- IUCN Globally Threatened
- Historical breeding population decline
- Breeding population decline
- Non-breeding population decline
- Breeding range decline
- Non-breeding range decline

WORDS – PETER COOPER | ART – RICHARD JARVIS

Nightingale

"The small birds whistle and the Nightingale sang" goes the old English folk song, testimony to the enduring role that our beloved Nightingales hold in our ancestral minds. This common lyric reveals an elevation of this bird's expressive sound as loftier, more eloquent, more emotional and human than any other songbird for the old folksingers of the past. And those who know his song will likely testify to this truth. There are other great singers out there but none can cast the deep spell that a Nightingale does. In the darkness of the thicket, these sonic masters create a meditative entrancement that has enchanted listeners for millennia.

The reams of literature, prose, poetry, mythology, folklore and folksong in devotion to the Nightingale are testament to how adored and powerful this little brown bird is. Nightingales are national symbols, popular music icons, and melancholic mascots for generations of artists. Yet, in recent decades his song has become perilously unknown in spite of his ubiquitous name, a demise reflected in a shocking population decline. Despite conservation efforts to restore habitats and encourage breeding the decline foretells a terrifying likelihood that we may be the last generations to have the privilege of listening to these birds.

2024 will mark the 100 years anniversary of the BBC's groundbreaking live outdoor broadcast, a radio first. And the two stars of that May 19th event were a Surrey Nightingale and the celebrated cellist Beatrice Harrison. Their live duet pioneered the idea of participatory artistic collaboration with nature. However, Harrison's collaboration was likely an innocent discovery, the cellist unaware of the legacy of musicians worldwide that have historically entered into conversation and ceremonial participation with these very human-friendly songsters.

That viral media stunt caught the nation's imagination, doing more to restore Nightingale appreciation than anything since. Nearly a century later my own 'Singing With Nightingales' concerts tread in her footprints as contemporary immersions, reintroducing Nightingales alongside musicians, to audiences reckoning with all the issues of nature access and connection. In these days of rapid nature depletion, the arts serve as vital tools in the conservation movement but are perhaps even more necessary as therapy for citizens facing the emotional weight of species loss. Such musical collaborations are acts of ornithological palliative care, offering a blueprint on how to show gratitude for what may not survive the oncoming decades of the Anthropocene. For to lose our Nightingales would be to lose a huge part of our collective identity.

WORDS – SAM LEE | **ART** – FLORA WALLACE

Whinchat

Your battered Mini was our escape from the city and student life, the opportunity to break free from the weekly routine of lectures and late nights. West from Exeter and accessing Dartmoor via Moretonhampstead, we'd head to the ruins south of Hookey Tor or track west up on to Challacombe Down. It was here that you showed me my first Whinchat. This was "your bird", you'd said. A restless migrant, easily overlooked. A bird of the margins.

Your totem became our totem that summer, the two of us inseparable as we criss-crossed the moor and its margins. Whinchats appeared here and there, much less often than the more familiar Stonechats, Wheatears and Meadow Pipits, but perhaps that is why they stood out.

It was late that summer that you handed me a cigarette card, picked up in a junk shop in Oakhampton and featuring a pair of Whinchats, facing one another along a branch, the male with a courtship offering. Everything about the image portrayed on the card was weird. Like a hand-coloured photograph that was slightly out of focus, the image jarred with the very real birds we had so often encountered. You apologised for this, but said these were "pocket Whinchats", birds that I could call up anywhere. It was a nice touch, and so typical of you. A thoughtful gesture. A gift I still have today. The card was old, older than we were then; older than you would ever be.

The following spring – your last spring – we saw our final Whinchat. You'd come to stay with us at the farm, and we'd taken a bus as far as Avebury. From there we'd walked up onto the downs at Fyfield and there, amazingly, we'd stumbled across a Whinchat. You suspected that this was a passage bird, something later confirmed by one of dad's books. Whinchat once bred on the downs here but no longer. This was a species that was being lost from many of its former haunts, a species that was slowly slipping away. I didn't realise it at the time, but the same could have been said of you.

You gravitated to these wild places for the same reasons as this bird, needing the solitude and sense of scale they offered. Increasingly, you struggled to cope with the realities and responsibilities of life, until one day, without me beside you, you shed those shackles on the edge of the moor. Even now, decades later, I have not been brave enough to venture back to our old haunts. I still hope of a chance encounter with a Whinchat, through which my connection to you may once more be restored.

CHANGING BOCC STATUS 1 2 3 4 5
RED-LIST CRITERIA APPLICABLE
- IUCN Globally Threatened
- Historical breeding population decline
- Breeding population decline
- Non-breeding population decline
- Breeding range decline
- Non-breeding range decline

WORDS – ELIZABETH PARKER | **ART** – BRIN EDWARDS

House Sparrow

The first time I saw a House Sparrow up close I was entranced. Who knew such a little brown bird had such intricate markings? I look at diagrams and read words like 'nape', mantle' and 'rump' (parts of a bird's body) and others like 'lesser coverts' 'tertials' and 'primaries' (types of feather). Both male and female House Sparrows look like they're wearing the sharpest of tweed suits, the male with his perfect combination of warm orange-infused browns, plus black and grey, and the female a more subtle (some books would say dull) palette of creams and browns. Imagine putting a series of browns together and coming up with something so beautiful. You couldn't do it. They're a work of art, House Sparrows.

And yet, they're still scruffy, which makes me love them even more. In my garden, they have an air of being a bit unwashed. Of having just got out of bed, they're crumpled, somehow. There's always a feather out of place, and their call – an almost constant "*Cheep! Cheep! Cheep! Cheep!*", which I translate as "*Hey! Hey! Hey! Hey!*" – makes them seem as though they're still a bit pissed from the night before. They have big parties in the soil (scientifically known as 'dust bathing'), which makes them look so ridiculous that I laugh and laugh and laugh.

They seem to do alright, here, because the houses are old and full of holes, and most of the gardens are – at the moment – still intact. But I worry about them. They have been lost from so many city streets as, one by one, houses have been 'done up' and gardens – front and back – have been paved over, or covered with decking, fake grass or some other version of DEATH, like stones or purple slate chippings, to make them low maintenance and easy to rent out. A paved, fake-turfed or otherwise dead garden is no place for a House Sparrow. They'll happily eat from your feeders but they need aphids and caterpillars to feed their young. And how many aphids and caterpillars did you last see on a pile of purple slate chippings?

They don't need much: old and holey houses to make their nests, long grass and a few native plants to provide aphids and caterpillars, plus a nice big hedge to hang out in at night. "*Hey! Hey! Hey! Hey!*" says the hedge, every day, at dusk and dawn. "*Hey! Hey! Hey! Hey!*"

Hey House Sparrows. Please don't go.

WORDS – KATE BRADBURY | **ART** – ESTHER TYSON

Tree Sparrow

I live with a view like one I grew up with. It's not the same, but it'll do. A village edge view of hilly farmland, where wind and the sun have recently joined the local harvests. Moving back north taught us the name 'Tree Sparrow', one with a seemingly permanent caveat of "they're in trouble" putting them on my list of dismissed things, that I'd probably never see.

Our rookie error of fastening a feeder to the bedroom window reminded us birds get up early and knock rather loudly. We stuck it where the sun doesn't shine, giving us an alternative point of view to the less and less watched television. One evening we came home to a beautiful outline of dust on the window, above the little feeder. The shape of flight, of legs outstretched, wings high, the size of an honest newspaper's front page. We didn't have the camera trickery to photograph it. But I tried and showed a friend who said 'Sparrowhawk'. And that's where it all started to interest me.

I borrowed a too-heavy, overly-complicated book, and began perching by the window looking at the little lawn. That first day, I saw a crack team of birds working among the uncut grasses, finding food. The big black book had an illustration that not only matched the birds but also their exact behaviour, position and concentration. Tree Sparrows. I'm partly colour blind so have no idea what you might see, but they look very sharp in their matching uniforms and caps. I've only seen them fighting once, like a Tarantino movie on the cobbles, but that was in Yorkshire and things might be different there.

I'd always thought they were House Sparrows, as "Tree Sparrows were rare, weren't they?" We quickly realised the dozens we half-saw daily were always Tree Sparrows, only ever been Tree Sparrows. Since then I watch them every day, see them learn to feed, flutter and fascinate. Someone who 'knew about these things' didn't believe me, or my luck. The local bird club had never had any records for our 'square', never mind a record of a breeding and thriving population. They gave me a spreadsheet which I must confess I have never filled in. I don't know why. Or what I'm waiting for. I'm still learning too.

But they're in the right place, among people they can tolerate, and a landscape that changes slowly. Personally, I believe they are here because of a 77 year old woman who has spent half her weekly shopping money on bird food for longer than I've known her. They make her, me, the Sparrowhawks, the school's hedges, and the village a very happy place to be indeed.

WORDS – JAMIE NORMINGTON | **ART** – EMMA PRICE

Tree Pipit

I'm above the beck in Naddle Forest, soaking up dappled May sunshine with my family after what feels like a long and brutal winter. Three months before, we planted a pair of Oaks to celebrate the weddings of two close friends. One is stretching elegantly skywards, the other sprawling and ungainly. Both are spring-flushed with green leaves and encircled by Bluebells. While the kids eat a snack cradled in the cleft of an ancient Alder, the decadent blossom of a Crab Apple lures me up the slope into the forest.

Escaping the white noise of the water, the forest's birdsong returns. A Tree Pipit, recently home from Africa, is using a high branch as a springboard. Silent as he ascends, his song as he parachutes back down slows like a steam train pulling in.

Song aside, Tree Pipits are easily overlooked. To many they are indistinguishable from Meadow Pipits, firmly in the 'little brown job' camp, never to be a poster species for nature conservation. Even their Latin name, *Anthus trivialis*, seems to render them unimportant, but there's nothing trivial about what the decline of Tree Pipits says about the British countryside.

Tree Pipits are birds of the in-between, of the messy and unproductive, the scrubby and chaotic. They like trees, but not too many. They like grazing, but not too much. Though they perch aloft, they nest on the ground. I came horribly close to crushing half a dozen of their brown eggs, loosely held in a grassy hollow, a few years ago. The loss of wood pastures, the grubbing out of hedges, the failure to replace the in-field trees that studded the countryside of old, the consolidation and intensification of farming, have all contributed to Tree Pipits ending up in such peril.

And yet, perhaps there is hope. Our wedding Oaks are among tens of thousands that my colleagues and I have planted at Haweswater over the past decade. They are widely spaced, grown from local seed, maturing slowly in our high, damp corner of the Lake District. These new trees stand ready to replace their forebears as they succumb to the inevitability of age. But the Tree Pipits are impatient; they are quick to find the new plantings and their numbers, locally at least, are on the rise.

As I head back down to my family I wonder who will grow fastest, my children or our trees. I hope both will be nurtured by Tree Pipit song.

WORDS – LEE SCHOFIELD | ART – DENIS CHAVIGNY

Yellow Wagtail

A bird equally at home as a sulphur-yellow streak in Britain's green fields of winter wheat or potatoes, as amid the grasses and Zebu cattle of West Africa's Sahelian wetlands. The Yellow Wagtail is a species I have been lucky enough to study at both ends of its flyway and, as for all the millions of birds that migrate between Europe and Africa, I remain in awe of how such a delicate-looking bird can complete this twice-yearly feat of endurance and navigation across one of the most inhospitable places on earth, the Sahara Desert.

Birds breeding in arable croplands often forage on bare ground, along tram lines or along ditches and tracks at the field-edge, or take invertebrates in flight from perches on the crown of the crop. They can shift nesting habitat and diet to enable them to secure multiple broods in a landscape that experiences dramatic seasonal changes in both vegetation and invertebrates. As winter wheat becomes tall and dense Yellow Wagtails switch to more 'open' crops like potatoes and field beans, and from feeding on flies and beetles to damselflies, although the increasing predominance of winter wheat may now limit the availability of late summer nesting habitat.

Much about Yellow Wagtail migration routes and their non-breeding ecology remains a mystery. Their leapfrog migration, where the most northerly breeders winter furthest south, 'leaping over' birds in-between, reflects a balance between the costs associated with migration and the optimal timing of breeding. Birds breeding in the north, where spring is later, can afford to delay migration, take advantage of the 'rain-related' surge in food availability to fatten quickly and still arrive on their breeding grounds in good time. Southerly breeders using such distant sites would have to fatten before the flush in prey or delay migration and arrive late on the breeding grounds, something known to reduce breeding success.

My field journal, from 1993, reminds me how days in the Sahel awaken to "the rhythmic sounds of Fulani women pounding millet, the calls of Venaceous and Red-eyed Doves and snapping twigs as goats browse the scrubby Acacia." A far cry from huge tractors, sprayers and combine harvesters lumbering into action on the dark peaty soils of the East Anglian Fens. But the ultimate driver of change in these two very different landscapes remains the same – relentless human pressure to meet the ever increasing demand for food and fuel. Certainly, when I first encountered Yellow Wagtails on their wintering grounds 20 years ago, I had no idea how much tougher life would become for this deceptively fragile-looking long-distance traveller.

CHANGING BOCC STATUS 1 - 2 - 3 - 4 - 5
RED-LIST CRITERIA APPLICABLE
- IUCN Globally Threatened
- Historical breeding population decline
- Breeding population decline
- Non-breeding population decline
- Breeding range decline
- Non-breeding range decline

WORDS – JULIET VICKERY | **ART** – PAUL THORBY

Hawfinch

After it was ringed, I was given the privilege of releasing the Hawfinch. But he had got hold of the bit of skin between my thumb and forefinger and was not inclined to let go. I was able to feel for myself the power of that enormous beak. When he did decide to release me there was a good deal of blood. "Job done!" he might have been saying to himself in a satisfied manner, as he flew off into the trees.

Hawfinch numbers in the UK are estimated to have declined by 76% since 1970, and we now have fewer than 1,000 breeding pairs. No one is quite sure why this is. I was lucky enough to be working with Will Kirby and the RSPB team investigating the reasons for the decline, making a film about the work for BBC Springwatch. Hawfinches are extremely secretive and nest right at the top of tall trees   the average nest height is 16 m. Will and the team had had the brilliant idea of fitting small cameras above the nests to try to discover why they failed.

The first challenge was to find the nests; tiny radio transmitters were attached to breeding females, caught nearby, which then led the team to the right trees. The next challenge was to climb each tree so that the miniature camera could be placed above the nest. By a useful quirk of fate, I am a qualified Rope Access Technician so I was able to help by climbing up to the top of one tree, finding the nest and fitting the camera. What a fantastic privilege to see the tiny white fluffy Hawfinch chicks, all gaping at me and hoping for a snack (sorry chicks). Camera in place, I dropped back down to the ground, all the while marvelling at my good fortune, both to see a Hawfinch nest and having been able to help in a (very) small way with the scientific research.

The cameras showed the nests were predated by Great Spotted Woodpeckers, Goshawks, Jays and Carrion Crows (but not once by Grey Squirrels, all you squirrel knockers out there!) but the percentage surviving through to fledging was, at 36%, within the range for population stability, so predation doesn't seem to be the main problem. The research is now exploring other avenues and I sincerely hope they can come up with answers to inform conservation efforts, so more people in the UK can catch the occasional glimpse of the extremely handsome, increasingly elusive Hawfinch.

WORDS – MARTIN HUGHES-GAMES | **ART** – SCOTT MAYSON

Greenfinch

As a child, I would marvel at books depicting colourful birds from far away lands. Staring at the impossibly beautiful and intense plumage of parrots, birds of paradise and toucans, I would wistfully wonder why the birds in my garden did not invoke the same feelings of awe. Don't get me wrong, the Blue Tits and House Sparrows that visited my childhood garden were full of personality, but they were familiar and, to a child, familiarity is akin to apathy.

Then one day, whilst charting my sightings to discuss later with my father, I saw a flash of emerald green within our hedgerow. I froze, knowing any movement would frighten off the bird before I could identify it. My patience paid off; a pair of Greenfinches emerged timidly from the hedge and tentatively made their way to our birdfeeder. One was more vividly green and yellow than the other. As they fed, they appeared to have little arguments over who would get prime position on the feeder, but eventually they settled and took their fill. Slack-jawed, I rushed inside to tell my father I had seen a bird, right here in our garden, that could easily rival any of its tropical cousins.

Now, I am lucky enough to have Greenfinches visit my garden regularly, but in their case familiarity has not turned to apathy. Alongside the iconic green/yellow feathers that seem to shimmer as they catch the sun come wonderful personalities, short sharp bursts of chatter and the flurry of beating wings as they jostle for position on my feeders.

I have come to realise that I am exceptionally privileged to have these birds in my garden. Sadly they face a danger that threatens to wipe them out not just from my garden, but from every garden across the country. A disease called trichomonosis is responsible for an unprecedented decline in Greenfinch numbers. It is transmitted via contaminated surfaces, including bird feeders and bird baths, and prevents the bird from feeding properly.

The good news is we can all do something to help halt this decline. By regularly cleaning garden bird feeders and baths we can significantly reduce the numbers of birds infected and help populations recover. Having the joy of birds feed in your garden comes with a responsibility to keep them healthy, so that they continue to prove to us all that British garden birds can give any tropical bird a run for its money when it comes to colourful plumage.

CHANGING BOCC STATUS 1 2 3 4 5
RED-LIST CRITERIA APPLICABLE
- IUCN Globally Threatened
- Historical breeding population decline
- Breeding population decline
- Non-breeding population decline
- Breeding range decline
- Non-breeding range decline

WORDS – AMIR KHAN | **ART** – LAURA ANDREW

Twite

Scuttling and tiptoeing between the rocks and long grasses, praying not to be spotted by a ravenous Merlin, the inconspicuous Twite goes about its business, in the moorlands, highlands and coastal areas of the UK.

I have only had the pleasure of seeing these birds a handful of times. The first time was on the outrageously beautiful Sanna Bay on the Ardnamurchan Peninsula on the west coast of the Scottish Mainland, which is the most westerly point of Britain. I was helping lead a guided tour that day and was lagging behind as I had stopped to attend to my shoelaces. I suddenly heard the sound of a single shy morris dancer, ringing their bells overhead at speed. Looking to the sky I saw my first Twite; indeed, it was a whole flock heading north low over the dunes.

The next time I encountered them, I didn't even see the birds; they were high up on the crags near an elevated viewpoint overlooking Loch Na Keal on the Isle of Mull. I only heard their little out of tune bells ringing, but they were making use of the fact that their relatively colourless plumage allowed them to blend in brilliantly with their bleak early spring surroundings, before the landscape had leapt into green mode.

My most recent encounter with Twite was by far my favourite. In February I was helping on a guided wildlife tour on the Isle of Islay, and we were heading to a moorland RSPB reserve called the Oa. A very short walk from the car park we saw a haze of small birds bouncing off the barbed wire fence line, feeding on the ground for a short time, and then bouncing back up to the fence. Sounds idyllic, but this was February on a Scottish Island. The whole of the UK was experiencing two huge storms, with record breaking wind speeds in some areas; it was so cold, and tough to stand up sometimes.

The whole of our group was standing there in our six-plus layers of cosy warm clothing, and yet these dainty little birds persisted. There must have been nearly 200 feeding in the field. As soon as they landed on the ground they simply melted into the grass, and only showed themselves when they needed a quick breather on the fence lines. The males occasionally showed their pale pink rumps, a new observation for me. After this encounter, I have the upmost admiration for these unimposing little birds which are often overlooked because they don't bear all the colours of the rainbow, but they are a certainly a special bird to me.

CHANGING BOCC STATUS 1 2 **3** 4 **5**
RED-LIST CRITERIA APPLICABLE
- IUCN Globally Threatened
- Historical breeding population decline
- Breeding population decline
- Non-breeding population decline
- Breeding range decline
- Non-breeding range decline

WORDS – INDY KIEMEL GREENE | **ART** – DAVID WOOD

Linnet

From the shadows of the briar and brambles, a tinkling comes. Carried on the wind; some notes whispered, some jingling loudly in my ears. It's melodious and random. Chaos, yet carefully crafted. Listening carefully, the tinkles are sprinkled with cheeky little "*woo-hoos*". I reckon I can almost see a wry smile peeking from the scrub.

This belter of a song belongs to a bobby-dazzler of a brown job: the Linnet. I think it's got to be one of the most underrated birdsongs going. The twinkling notes, soft whistles and warm chattering of this little finch are simply joyful. I'd like to spread that song thickly on toast and guzzle it. It's a voice that never fails to uplift and comfort, like a hot brew. The song springs from the tangle of vegetation in front of me. Arcs of prickles; rose and blackberry. Binding tendrils of vetch and bryony. Stubborn saplings of birch and Rowan. I'm wandering through the scrub realm and it's deliciously messy. It's also the type of place you're likely to be blessed with Linnets and that magical song.

Scrubby areas are wonderfully exciting to me. Where there's mess, there's opportunity. Forgotten corners, untidy edges and bushy hedgerows are precisely where wild things like to dwell. And for me, the Linnet encapsulates the spirit of scrubby messiness perfectly. During lockdown, many of my regular nature reserves were too far away to justify visiting on my daily walk. Instead, I started bumbling down unexplored footpaths near the town I lived in – finding unexplored fields, industrial estates, woods and river floodplains. The humble, pink-bellied Linnet became a sort of lucky charm during these wanders. I learnt that where I heard them sing, other wildlife was never far away.

As I yomp towards this bramble patch, a Linnet cloud explodes from the green. A flock of finches, flowing through the sky; twitters, tinkles and "*woo-hoos*" galore! A few of them land in a birch tree nearby, and I get a chance for a really good ogle.

Linnets can be scoffed at in the birdwatching world. Banished to a heap of plain and ordinary brown jobs – barely worth a lift of your binoculars. But take a closer gander, and this little bird's spectrum of brown widens. I find using a different word for these colours tells a different story. Linnets aren't simply 'brown'. They contain marmalade and ginger. Rust and cinnamon. Salt-worn driftwood and Scots Pine bark at sunset. And then you get the males donning a soft pink cap and proud rose belly. I could up the romanticism more, but you get the gist. Yes, I think they're ordinary – but wonderfully, defiantly so.

CHANGING BOCC STATUS ①-②-③-④-⑤

RED-LIST CRITERIA APPLICABLE
- IUCN Globally Threatened
- Historical breeding population decline
- **Breeding population decline**
- Non-breeding population decline
- Breeding range decline
- Non-breeding range decline

WORDS – LUCY LAPWING | **ART** – JAMEY ANNE DOUGLAS

Redpoll

A balmy autumn had pursued a wet summer but today was suddenly bitterly nippy. November, yet a few Arctic and Common Terns having postponed their trip south, were still dive-bombing the strait. Gulls mewled, Oystercatchers piped. The lighthouse bell gave a mournful clang. Reminiscent of monasteries, it brought the monks to mind who had populated a clas across the water on Ynys Seiriol, predating the lighthouse (built in 1831 after the Rothsay Castle was shipwrecked) by more than 1,000 years. The pebbles were rain-spattered, the rocks etched and stippled. Valerian flowered. A kayaker paddled. The seasons it seemed, were adrift.

I turned inland, bound for Priordy Penmon; the priory which the clas had spawned. Y Fenai, gun-grey and impatient, nagged the stony shore to my left. Across it, mountains shouldered such cumbersome cloud banks that I thought at first it had already snowed. The mountains were hard and stern. Winter, they intimated, (and the ecosystems it sustained); winter too now was muddled, threatened, but they would fight for it while they still could. Either side of the toll road, small trees poked through scrub, underpinned by limestone and shale. A chill wind tugged at the bronze leaves still clinging to branches like verdigris plant labels. Some of the leaves were tumbling across the path, but not all of them, it transpired, were leaves.

Flocks of finches were passing over my head, many perhaps arriving from Scandinavia via the mainland; Chaffinch, Goldfinch, Brambling and Lesser Redpoll – llinos bengoch leiaf in Welsh. Some of these migrants were dropping into the scrub. It was the Lesser Redpolls I had mistaken for leaves – several of them were bouncing, chattering and tumbling across the path, bowling along like old fruit, adding colour to the wintry coast.

Lesser Redpolls are native to Britain and Europe, but have also been introduced to New Zealand. Their populations fluctuate with habitat, and are currently slumping in Britain. These small finches, stripy brown save for red heads and flushed breasts, feed on tiny wild seeds, and those of birch, larch and alder – nyjer too as it happens, prompting recent forays into Goldfinch-friendly gardens. They are often reported hanging out with Siskins in treetops, dangling like strange apples as they feed. But most of my encounters have been unexpected – on untidy paths and in scruffy young mountainous plantations – edgelands; which probably says as much about me, as it does about the importance of scrub.

WORDS – JULIE BROMINICKS | **ART** – MYLES MANSFIELD MA

Corn Bunting

Somehow the Corn Bunting has always been one step ahead of me. We've never been aligned, as current corporate parlance would have it. Childhood holidays in Wales featured seasoned locals who talked of the "jangly-bird" that used to sing from wires along the path that led down to the sea. "Haven't seen one for a while," they said, "but they must still be out there." Hmm. I looked every spring and summer, but the Corn Buntings appeared to have moved on. Twenty years later I came to live in a village in the Norfolk Brecks and on an early foray across nearby fields I heard one singing in the distance. "Great," I thought, "there are Corn Buntings almost within earshot of the garden!" It remains my one and only local record.

In truth of course, the recent history of this enigmatic farmland bird has been of steps taken backwards, rather than forwards. The Corn Bunting was not moving ahead of me, it was simply disappearing. The last quarter of the 20th century saw numbers in Britain shrink by as much as 90 per cent, with scores of local – and often rapid – extinctions, and formerly buoyant populations reduced to relict outposts. Once widespread and common across much of Britain, Corn Buntings always seemed to have been in a state of ebb and flow. One writer in the 1950s dismissed their distribution as "notoriously capricious". Invariably shifting, evading certainties and defying expectations.

Brown and dumpy, even a tad clumsy, the Corn Bunting presents a prosaic aspect. Certainly it has none of the glamour or pulling power of other species that have suffered free-fall declines. How to compete with the rakish banditry of the Red-backed Shrike and the quirky crypto-contortions of the Wryneck? Yet in their own way, Corn Buntings are avid attention-seekers. Pioneering bird photographer Emma Turner described the faux bravado of jousting males in their display flights, legs a-dangle, and how "after shivering a lance or two they return to their posts quite satisfied with themselves for the time being." Then there's the song. Repetitively shimmering, wheezing, key-jangling, a hazy hot weather sound. The only constant in a bird that now wanes far more than it waxes.

Undone by changes in agriculture (fewer spring-sown cereals mean less autumn/winter stubble, which in turn means less food is available for the mainly seed-eating adults) and the drenching of our farmland in agrochemicals (fewer insects to feed chicks), the Corn Bunting sits depleted and fragmented. Rattling away on a nearby strand of barbed wire. Or not, in my case at least.

WORDS – JAMES PARRY | **ART –** GREG POOLE

Cirl Bunting

It's more than 10 years since I moved to south Devon and began to explore the coast path. I've come to love the mild marine environment, the relative isolation and the drifts of colourful wildflowers growing by the path in spring and summer. Several people, though, told me to look out for the Cirl Bunting, a rare but distinctive bird, a local speciality.

I looked, and at first was rewarded with a few fleeting sightings. As I got to know the area, I discovered Cirl Bunting hot spots along the coast path. Here groups of the birds might skulk in the coastal scrub or a lone male might perch on the top of a hedge singing his heart out. With his striking black and yellow-striped head and neck and his olive-green breast band clearly visible, this was a joyous sight. I also learnt to recognise their characteristic metallic, rattling song.

In my voyage of discovery, I benefitted from the conservation efforts of the RSPB. The Cirl Bunting was once common across much of the southern half of the UK but declined dramatically in the latter part of the 20th century. By 1989, only 118 pairs remained, confined to coastal farmland in south Devon. The RSPB realised that changes in farming practice were responsible for this downward spiral. Agricultural intensification had driven a shift from spring-sown cereals to autumn-sown, so that few arable fields were left as weedy winter stubble with seed and spilt grain as food for the birds. Grubbing out hedges had taken away nest sites. Loss of unimproved grassland and increased pesticide use had reduced invertebrate numbers, summer food for the birds.

The RSPB worked with farmers in south Devon to support the birds by reinstating some traditional agricultural practices, underpinned by government agri-environment schemes. The result was a spectacular increase in Cirl Bunting numbers along the south Devon coast to over 1,000 pairs by 2016. More recently, the bird has spread to east Devon and a few have even reached Dorset. This is good news but the bird's survival depends on the continuing goodwill of farmers and government agri-environment support.

I still walk the coast path and sometimes I pause quietly in a meadow, surrounded by wildflowers. Butterflies flicker around me and the metallic, rattling song of the Cirl Bunting echoes from coastal hills. It's a pleasure, but one tinged with sadness: much of our countryside used to be as rich as this, but sadly no longer.

WORDS – PHILIP STRANGE | **ART** – ADRIAN DUTT

Yellowhammer

The sun is bright in the sky above. There's not a cloud to be seen, and in the air there hangs the promise of a glorious summer's day. We're here to see orchids, but can't help being distracted by other wildlife; the reserve is filled with the song of birds and the buzzing of bees. Butterflies swoop and dance over wildflowers that nod their heads gently in the whisper of a breeze. As we walk to a spot that is meant to be a hot spot for the orchids, our eyes catch a splendid yellow flash amongst the lush green of a tree. Binoculars fly up, and agreement is reached – a male Yellowhammer.

About the size of a sparrow, he sits confidently in the tree, seemingly unfussed by our regard. Facing us, we can clearly see his yellow head and underparts and he seems to shine in the bright sun. A moment later, our identification is confirmed as he opens his beak to sing "*a little bit of bread and no cheese*".

Until I read Dr Mya-Rose Craig's entry in *Red Sixty Seven*, I hadn't clocked that Yellowhammers could be flocking birds in winter, joining other buntings and other small passerines. Reaching back to trawl through my memories, I can only recall seeing one or two individuals at a time. Maybe I've seen a small flock together and I just can't remember it. I wonder, is this just a case of bad luck on my part, or another negative example of shifting baseline syndrome? I speak with a few fellow naturalists of a similar age, and just one says they've seen a flock, so it seems it might be the latter.

If it is such a case, I ought not to be surprised. After all, there's a reason that the Yellowhammer is included in this book – it's not doing well. This little yellow bird has been in rapid decline in the UK, from Green List status in the first Birds of Conservation Concern assessment, then skipping straight over Amber to Red in the next and all the following assessments. From the mid 1980s to now, the last few decades have not been kind to the species – a sea of downward trending graphs on the BTO website is a depressing view.

I hope there's change afoot for this sweet songbird, and that the baseline shifts in the opposite direction. Perhaps future generations will read this entry and ask what on earth I'm on about; there are Yellowhammers everywhere. But gloomily I ask myself, how likely is that?

CHANGING BOCC STATUS 1-2-3-4-5
RED-LIST CRITERIA APPLICABLE
- IUCN Globally Threatened
- Historical breeding population decline
- Breeding population decline
- Non-breeding population decline
- Breeding range decline
- Non-breeding range decline

WORDS – MEGAN SHERSBY | **ART** – ALISON DEEGAN

In memoriam

Since the last Birds of Conservation Concern review, carried out in 2015, the Golden Oriole has now been designated as being extinct as a breeding species in the United Kingdom

It would be fair to say that the Golden Oriole has never had more than a toehold in the UK as a breeding species, its population thought to be linked to populations in the Low Countries and, because of this, never self-sustaining. The last breeding record was in 2009, from Suffolk; prior to this, the last confirmed breeding outside of the county was back in 2003.

The species remains a scarce passage visitor, typically recorded from coastal localities in the south-east of the UK during spring. Research, exploring the potential future distribution of birds under climate change projections, suggests that the Golden Oriole could expand its current breeding range into south-east England, though the poor breeding success of English nests, linked to poor weather in June, may make successful recolonisation less likely. We shall have to wait and see.

WORDS – MIKE TOMS | **ART** – WILL ROSE

Our contributors

LAURA ANDREW is a wildlife artist whose paintings are inspired by the natural world, particularly British birds. Her work has been shown throughout the UK including at the SWLA's annual exhibitions. www.lauraandrew.com

MARK ANDREWS is a wildlife artist and tour leader, with a passion for birds and mammals, especially wild cats. Twitter: @BeidaiheBirder

GRAHAM APPLETON, former BTO Director of Communications and author of the popular WaderTales blog series, is a life-long advocate for wader research and conservation. Twitter: @GrahamFAppleton

HOLLY ASTLE is an illustrator based in Cornwall. She takes inspiration from the natural world around her and is passionate about its preservation. www.hollyastle.co.uk

JEFF BAKER worked at BTO for 47 years, initially in the Ringing Unit before heading up membership and then marketing before retiring in 2017. A keen artist, he has illustrated several books.

DAWN BALMER is a life-long birdwatcher and nature enthusiast. She is Head of Surveys at BTO, Chair of RBBP and member of the British Birds Rarities Committee Twitter: @debalmer

DAVID BENNETT, is a full time professional wildlife artist and illustrator, creating lively representational and intimately observed work, captured through field sketches and paintings of travels around the UK.

BO BEOLENS is a lifelong birder, champion of disability access, author, columnist and currently treasurer of Sandwich Bay Bird Observatory Trust, but best known for the online resource he created www.fatbirder.com

TONY BLYTHE is a self taught illustrator working mainly in ink and watercolour and uses a loose line style and vibrant colours to inject energy and life into his work. Twitter: @itsnotaboutwork

TESSA BOASE uncovered the surprising story of the RSPB's female founders for her book, *Etta Lemon: The Woman Who Saved the Birds*. She lives on the Sussex coast.

KATHERINE BOOTH JONES is a BTO scientist with a background in seabird ecology. Her love of seabirds was kindled by the charismatic Herring Gulls of her hometown of Hastings.

KATE BRADBURY writes about and presents programmes on gardens and nature, A patron for Froglife and the Bumblebee Conservation Trust, Kate is also an ambassador for the Royal Horticulural Society. Twitter: @Kate_Bradbury

BRIAN BRIGGS is a nature reserve manager for the Wildfowl and Wetlands Trust and frontman of the band Stornoway, who's latest album *Bonxie* features the songs of twenty different species of birds. www.stornowayband.com

JULIE BROMINICKS writes mostly about the Welsh landscape for BBC Countryfile Magazine. Her first book, *The Edge of Cymru; A Journey*, was published by Seren in autumn 2022.

ADAM BUXTON is a British writer, comedian and podcaster.

STEVE CALE is a wildlife artist and ornithologist, whose work has appeared in numerous publications. He has undertaken work for many conservation organisations. His paintings are owned Internationally. wwwsteve-cale-artist.co.uk

HAYLEY CHAN is a British Chinese illustrator and graphic designer, whose work is about the natural world, stories, environmental & social good and culture.
Instagram: @huetone_illustrator

DENIS CHAVIGNY is a wildlife artist who works from field sketches and has spent the past 30 years on the banks of the Loire. An ornithologist first and foremost, birds and their future continue to occupy his life.

FAITH CHEVANNES is a printmaker from Cornwall who has a passion for wild birds. She exhibits her work nationally and has been included in the Royal Academy Summer Exhibition. www.faithchevannes.com, Instagram: @faithchevannesart

MILES CLUFF is a keen young birder and recent chemical engineering graduate with a passion for wildlife art, currently working with the BirdGuides news team, as well as an ambassador for Opticron. Twitter: @miles_cluff

DANIEL COLE SWLA, is a landscape and bird artist based in Cornwall; his work is in the collection of the Royal Cornwall Museum and in 2016 he won Swarokski Bird Artist of the Year.

PETE COOPER is a wildlife conservationist working on reintroducing native species such as Glow-worms and Harvest Mice and rewilding for the Derek Gow Consultancy.
Twitter: @PeteMRCooper

MACKENZIE CROOK is a multi-award winning actor, writer and director, who works across theatre, television and film. Jerusalem, Detectorists, Pirates of the Caribbean, The Office. www.mackenziecrook.com

NICOLA DAVIES has written over 80 books for children and adults, fiction, non fiction, poetry and picture books. Her latest book is an environmental thriller *The Song That Sings Us*. Twitter: @nicolakidsbooks

ALISON DEEGAN is a printmaker and an archaeologist who strives to capture the textures and colours of things with wings and places with history in carved and etched lino prints. Twitter: @a_deegan

HEATHER DEVEY is a North East based naturalist, she is Founder and Co-Director of Wild Intrigue, a social enterprise created to "inspire, educate and rewild". www.wildintrigue.co.uk

JAMEY DOUGLAS is a Suffolk-based artist, who specialises in painting and drawing UK wildlife (with a particular interest in birds). She uses watercolour and ink to capture the beauty of the natural world and bring her subjects to life. Instagram: @jameydraws

AARON DUFF (aka Hector Gannet) is a singer songwriter from North Shields who works solo and with his band under the *Hector Gannet* moniker.

ADRIAN DUTT is an illustrator, printmaker and musician based in North Somerset. He uses field sketches, photography and linocuts to explore his connection to the bird world. Instagram: @howlingkestrel Twitter: @adriandutt

DR MARK EATON is an ornithologist and conservation scientist who, after two decades with the RSPB, now works as Secretary of the Rare Breeding Birds Panel and chairs the European Bird Census Council. Twitter: @Mark_A_Eaton

BRIN EDWARDS is a wildlife painter specialising in birds with attention to light, colour and some elements of abstraction. www.brin-edwards.com

ROSIE ELLIS previously worked for The National Trust, with Cornwall Wildlife Trust and in Cape Verde. Sailing the Atlantic and completing a master's at The University of Reading between roles. She is now the warden of Lundy Island. Twitter: @lundylandmark

LISSA EVANS has written novels for both adults and children, including the best-selling *Old Baggage*, and *Their Finest*, which was adapted into a much-loved film. lissaevans.com Twitter: @LissaKEvans

JOHN FANSHAWE is an author, curator, and environmentalist, who works for BirdLife and the Cambridge Conservation Initiative. He co-founded New Networks for Nature in 2009, and is based in north Cornwall.

DR ANNETTE FAYET is a researcher in seabird ecology, currently based at the Norwegian Institute for Nature Research. She specialises in tracking seabirds to uncover their movements at sea. Twitter: @AnnetteFayet

JAKE FIENNES has a 30-year career in land management and is Director of Conservation for the Holkham Estate and Holkham National Nature Reserve on the North Norfolk coast. He is author of *Land Healer* published in 2022. Twitter: @jake_fiennes

JONNIE FISK enjoys the company of the Humber estuary's Brent Geese, Blackcaps and brick red Bar-tailed Godwits; living vicariously through their journeys and those of similar avian migrants. Twitter: @jonnie_fisk

BEATRICE FORSHALL is an artist whose work draws upon the natural world, and her printmaking revolves around species and themes central to conservation.
www.beatriceforshall.com

JESS FRENCH is a veterinarian, zoologist, author and broadcaster. She specialises in writing for young and family audiences on the themes of nature and conservation. Twitter: @Zoologist_Jess

FEDERICO GEMMA is an Italian wildlife artist and member of the Society of Wildlife Artists. He has made illustrations for calendars, books, magazines and other publications including a set of bird songs stamps for the Royal Mail. Twitter: @Federico_Gemma

JOSIE GEORGE is a writer and visual artist, Guardian Country Diarist, and author of the highly acclaimed *A Still Life: a memoir*. Find her on Twitter and Instagram as @porridgebrain

JENNIFER GILL is Professor of Applied Ecology at the University of East Anglia and studies the migration ecology and conservation of waders, working with researchers and volunteers across Europe Twitter: @jengill3

DAVID GRAY, Brit and Grammy nominated songwriter, has released twelve albums in his thirty year career, including his breakout multi-platinum hit *White Ladder*, which went on to sell seven million copies.

JAMES GREEN is a Sheffield-based artist and printmaker who specialises in linocut and screenprint. His work generally revolves around landscapes, UK wildlife and surreal donkey compositions. Instagram: @jamesgreenprintworks

MALCOM GREEN is a storyteller, experienced educator and workshop leader. He teaches at Newcastle University, and works in collaboration with scientists to create stories that bring their work to a wider audience. www.malcolm-green.co.uk

INDY KIEMEL GREENE is a 17 year old naturalist and volunteer with the RSPB in Sherwood Forest. Indy has been fortunate enough to have featured on Springwatch talking about Goshawks and Lesser Spotted Woodpeckers. Twitter: @GreeneIndy

MARK GURNEY is a naturalist and digital illustrator, happiest when birding. Twitter: @MarkGurn

DAVID HALL is a Northumbrian artist/illustrator and writer inspired by the natural world, producing uniquely stylised graphic images for prints and self-published books. www.davidhallartist.info

PAUL HARFLEET'S #BirdsCanFly is an ongoing body of work that explores the history and colonialism of ornithology from a queer perspective. His 'gentle references' use fashion, make-up and styling to depict the birds he illustrates that reflect the aesthetic nature of his subjects and explore their cultural context and history. Twitter: @ThePansyProject

PHILIP HARRIS is an Illustrator and printmaker currently based in Devon. He works prominently in the medium of dip pen and ink which he inherited from his Grandfather. Twitter: @PhilipHarrisArt

CELIA HART is an artist, printmaker and illustrator based in Suffolk, her main inspiration is the landscape and wildlife she encounters on walks through the seasons. www.celiahart.co.uk

ANDREW HASLEN is a Suffolk-based wildlife artist, whose work is inspired by the landscape and wildlife that surrounds his home. His work has been published in several books, most recently *The Art of Gamehawking*. www.andrewhaslen.com

REN HATHWAY is a wildlife artist. He studied fine art and is best known for *Thrushes of The World* and his illustrations in the *Isles of Scilly Bird Group Natural History Review*.

JOHN HATTON is an artist producing work inspired by direct observation, his linocuts are printed using an Albion press built in 1884 and weighing 1.2 tonnes, he is an elected member of the SWLA. jhattonart@gmail.com

ALICIA HAYDEN is a writer, filmmaker, and wildlife artist – she published her debut poetry book *Rain before Rainbows* in 2020, and has received recognition from DSWF for her wildlife artivism. Twitter: @aliciahaydenwildlife

NICK HAYES is an illustrator, author of *The Book of Trespass* and co-founder of the Right to Roam campaign. He lives on a barge on the Thames. Instagram: nickhayesillustration

LISA HOOPER (ASWLA) is a wildlife artist living and working in south-west Scotland. She exhibits widely throughout the UK and her work has been published in books and greetings cards. www.hoopoeprints.co.uk

ALEX HORNE is a comedian, writer and son of a birder. He's written books about words, birds and puzzles, comes up with the tasks for Taskmaster and cares for every one of Greg Davies' whims. Without exception. Twitter: @AlexHorne

MARTIN HUGHES-GAMES is a zoologist and television presenter (BBC Springwatch and other programmes). He produced television programmes for 30 years and wrote an amusing book about it titled *A Wild Life*. Twitter: @MartinHGames

LIZ HUMPHREYS is a self-confessed lover of seabirds and works for the BTO as a Principal Ecologist. You can find her on Twitter as @KittiwakeGirl.

ANNIE IRVING is a dedicated biodiversity recorder, an avid walker, an enthusiastic birder and butterflier, who aspires to engage people with the natural world through her daily blog: earthstar.blog.

RICHARD JARVIS is a Leicestershire based printmaker, his work has been exhibited widely and reproduced in several books and as cards. He was elected a full member of the SWLA in 2019. www.richardjarvis.me.uk

KIT JEWITT is a Geordie birder, hare-brained schemer and fundraiser for bird conservation. Host of the Golden Grenades podcast. Twitter: @YOLOBirder

EMILY JOÁCHIM is an owl ecologist, bird ringer and nature writer. She has worked in nature conservation for 14 years with a special interest in nocturnal wildlife.
Twitter: @emilyjoachim www.littleowlproject.uk

SHAUNA LAUREL JONES writes about the environment, art, and identity, and is an Icelandic-to-English translator. She was shortlisted for the 2021 Nan Shepherd Prize for underrepresented voices in nature writing. www.shaunalaureljones.com

KABIR KAUL is an award-winning conservationist, campaigning for London's wildlife through writing, broadcasting, public speaking and social media. He served on the Mayor of London's Rewilding Taskforce in 2022. Twitter: @Kaulofthewilduk

DR AMIR KHAN is an NHS GP, Sunday Times Bestselling author and TV presenter. He is Vice President of the Wildlife Trusts and works closely with the RSPB, the Hedgehog Society and Butterfly Conservation. Twitter: @DrAmirKhanGP

MIKE LANGMAN is a lifelong birder – particularly seawatching, an avid field sketcher, book and magazine illustrator, bird and wildlife guide, and chair of a local conservation group in Torbay. Twitter: @clennonvalley

LUCY LAPWING is a casual naturalist, wildlife communicator and proud nature nerd. She currently campaigns on wildlife issues, as well as sharing her wild finds and passion for them on social media, @Lucy_Lapwing

SAM LEE plays a unique role in the British music scene. He is a highly inventive and original singer, folk song interpreter, passionate conservationist, song collector and successful creator of live events. www.samleesong.co.uk

OLIVIER LEGER is a marine wildlife artist. He draws inspiration for his intricately detailed drawings from his experiences underwater and his passion for conservation.
Twitter: @O_Leger www.olivierleger.com

M. G. LEONARD is the bestselling children's author of *The Twitchers* series, about a gang of crime solving birders. Her other books include *Beetle Boy* and the *Adventures on Trains* series. Twitter: @MGLnrd

NORA LIGUS combines art and nature in paintings and taxidermy. In a former life she studied costume design, but nowadays you can most often find her in nature with her dog.
Twitter: @greyauk

STEVEN LOVATT is the author of *Birdsong in a Time of Silence* and editor of *An Open Door: New Travel Writing for a Precarious Century.*

RICHARD MABEY has been a pioneering voice in modern nature writing for more than five deacdes. He has consistently explored new ways of thinking about nature and its relation to our lives. www.richardmabey.co.uk

BENEDICT MACDONALD is a writer and conservationist, and the author of *Rebirding*, *Orchard: A Year in England's Eden* and *Cornerstones*. Having worked for 10 years on wildlife documentaries, he now works as Head of Nature Restoration for Real Wild Estates, turning large areas towards ecological restoration. Twitter: @Rebirding1

ANDREW MACKAY is a freelance artist and illustrator whose work has appeared in many publications. He works in a variety of media including acrylic, linocut, coloured pencil and digital. www.ajm-wildlife-art.co.uk

MYLES MANSFIELD was compelled to start painting 15 years ago. He has since sold his paintings and sculptures worldwide, and been selected for many national and international exhibitions. www.mylesmansfieldartist.co.uk

MELANIE MASCARENHAS uses direct, observational fieldwork and the interplay of colour, movement & sound drawing to convey the rhythm and energy of a place and its wildlife, through expressive painting, drawing and printmaking. Instagram: @melaniemascarenhas1

SCOTT MAYSON is the BirdTrack Organiser and co-founder of Alula designs. An active suffolk-based birder and secretary of the Suffolk Ornithological Records Committee. Twitter: @suffolkbirder

MEGAN McCUBBIN is a zoologist, author and TV Presenter working across BBC Springwatch, ITV's This Morning and CBBC's Planet Defenders. Twitter: @MeganMcCubbin

JAMES McCALLUM is an artist based in North Norfolk. He is a graduate of The Royal College of Art and is best known for his watercolour paintings of the natural world, particularly birds, made outdoors from life. www.jamesmccallum.co.uk

TOM McKINNEY is a classical music broadcaster who presents BBC Radio 3's most popular programme Essential Classics. His bird notebooks go back to 1989 and his favourite places to go birding are Cornish headlands and Spurn Point. Twitter: @tom_mckinney

HARRIET MEAD is current president of the Society of Wildlife Artists, and makes her sculptures out of old tools and scrap. She has worked with various NGOs on projects highlighting conservation stories. www.harrietmead.co.uk

JIM MOIR has been making paintings, drawings and prints since the early 1980's. This practice has continued alongside acting and comedy for which he is better known, although to him they are all forms of artistic expression. www.vicreeves.tv

GARY MOORE is a zoologist and professional wildlife sound recordist, who has worked on many high end wildlife documentaries from around the world. He has a large database of bird songs and calls from all corners of the UK. Twitter: @mooresounds

CHARLIE MOORES is a campaigner and birder with a particular abhorrence of wildlife crime. He writes and podcasts at Off the Leash. Twitter: @charliemoores

HELEN MUSSELWHITE is an illustrator specialising in paper craft. She lives and works on the edgelands of deepest South Manchester. www.helenmusselwhite.com

DAVID NOBLE is the BTO's Principal Ecologist for Monitoring, previously having worked on seabirds in Canada and cuckoos in Africa. He is particularly interested in causes of change, using citizen science, and has a soft spot for non-natives. Twitter: @aparanoble

JAMIE NORMINGTON walked 200 miles across England to fundraise for hundreds of nature books for school libraries. He learns, laughs and shares daily, often whilst working for Cumbria Wildlife Trust. Twitter: @tlwforcumbria

KATHRYN O'KELL has been making carvings for over 30 years. Her work appears in public and private collections worldwide and she exhibits regularly in selected galleries around the country.

ELIZABETH PARKER is a writer and occasional poet, from a farming background, rooted in a landscape dominated by the Marlborough Downs and Savernake Forest. Twitter: @WindChatterer

JAMES PARRY is a writer on wildlife, heritage and art. He has published extensively on these subjects and has a particular interest in the history of bird photography. www.jamesvparry.com

PETER PARTINGTON, wildlife artist, has written and illustrated many books, and taken part in many conservation projects world-wide. He most enjoys being out in the Suffolk landscape with his sketchbook. www.thewildlifeartist.co.uk

RUTH PEACEY is a naturalist, birdwatcher and freelance conservation/wildlife film maker. She has worked on a number of BBC documentary series, including Natural World, Springwatch, Life in the Air and Planet Earth II. Twitter: @ruthpeacey

NIK POLLARD is an artist / illustrator focussing on the natural world. As a member of the Society of Wildlife Artists he has contributed to several international conservation projects. Instagram: @nik.pollard

KARINE POLWART is a multi-award-winning Scottish singer, composer, theatre-maker, storyteller and writer. Much of Karine's music and writing is steeped in place, hidden histories, scientific curiosity and folklore. Twitter: @IAMKP

TIM POND is passionate about the natural world and education, and prefers to draw directly from nature. He has worked as an expedition artist in Alaska and sketched leaves in a Mexican tropical forest. Twitter: @timothypond

GREG POOLE (1960–2018), artist and naturalist, was respected for his depth of knowledge and expressive visual responses to the natural world. He contributed to many international conservation projects.

JIM PRATT, born in London in the Blitz, worked as a forester and forest pathologist for 50 years throughout Britain, Europe and N America. Retired, he now lives in Scotland.

EMMA PRICE is an amateur artist from Hartlepool in north-east England, with a special interest in birds, nature and wild landscapes. Her online gallery can be viewed at www.painters-online.co.uk/artists/ep55

EILEEN REES is a researcher and conservationist, particularly of swans and other waterbirds. Following over 40 years at the WWT, she continues to Chair the IUCN-SSC Swan Specialist Group in supposed 'retirement'. Twitter: @EileenRees

SARA RHYS is a mixed media artist, painter, and illustrator of several children's books. www.sararhys.co.uk

LAURENCE ROSE is a conservationist, researcher and writer. He retired from the RSPB in 2022 after 38 years, and now works on the cultural and ethical dimensions of conservation. Twitter: @LaurenceR_write

WILL ROSE is a British animator, illustrator and wildlife fan. He currently creates animations for 'Yoga with Adriene' and has worked as animator and designer on projects with David Attenborough, Hey Duggee and Peppa Pig. He also makes his own award winning animated short films. www.willswork.co.uk

BEN ROTHERY is a detail-obsessed illustrator and writer from Norwich, via Cape Town who likes pencils and heavy objects. He combines multiple processes to create intricate and delicate illustrations, full of fine detail and vibrant colour. www.benrotheryillustrator.co.uk

DAN ROUSE is an author and presenter from Wales featuring on BBC Radio Wales, Countryfile and S4C. She's passionate about the environment and sustainability, frequently writing articles and blogs on the subjects. Twitter: @DanERouse

STEPHEN RUTT is a birder and author. He has written three award-winning books about birds, nature and landscape. He lives in Dumfries with his wife, baby and cat. Twitter: @steverutt

LEE SCHOFIELD is Senior Site Manager for the RSPB at Haweswater in the Lake District, and author of *Wild Fell: Fighting for Nature on a Lake District Hill Farm*. Twitter: @leeinthelakes

JOSÉ ANTONIO SENCIANES is a watercolorist specialising in birds and landscapes, with more than 20 years of experience in wildlife art. Twitter: @jasencianes

STEWART SEXTON takes photographs but isn't a photographer; he paints but is not an artist either. You can find him on Twitter: @Stewchat

MEGAN SHERSBY is a naturalist, writer and often-bumbling gardener based in Northamptonshire, who has worked for BBC Wildlife Magazine, the National Trust and the Wildlife Trusts. Twitter: @MeganShersby

CELIA SMITH is an artist who uses recycled wire to make sculptures and drawings of native British birds. Instagram: @celiasmith_wireartist.

KEN SMITH had a long career with the RSPB doing research on threatened species and habitats. **LINDA SMITH** was a civil servant developing controls on chemicals and dangerous substances in the environment. They are both now enjoying retirement in pursuit of many ornithological projects including initiatives to help the study and conservation of the Lesser Spotted Woodpecker. Twitter: @LesserSpotNet

MANUEL SOSA is a Spanish illustrator and painter, whose work focusses on Iberian nature, with a realistic pictorial style and advanced light and colour work. Biologist and musician too, his works dress the walls and hundreds of publishers' pages around the world.

BECCY SPEIGHT is the Chief Executive of the Royal Society for the Protection of Birds (RSPB) and previously led the work of the Woodland Trust. Twitter: @BeccyRSPB

KATIE STACEY is a nature writer who lives in Spain with her partner Luke Massey, and their two sons, at Wild Finca their agriwilding project. Twitter @wildfinca

PAUL STANCLIFFE is a lifelong birder, has worked for the BTO in communications, co-authored two bird guides and was the resident bird recorder for the Isles of Scilly between 1999 and 2004.

DAVID STEEL has lived, worked and managed seabird islands for 22 years including the Farne islands (2001–2014) and Isle of May (2015–present). Twitter: @Steelyseabirder

TOM STEWART writes about birds, music, people and places. Born in Manchester, he works for BTO and lives in Norfolk.

PHILIP STRANGE is a writer, scientist and naturalist who lives in south Devon. https://philipstrange.wordpress.com

BRIGIT STRAWBRIDGE is a wildlife gardener and amateur naturalist, who writes and speaks to raise awareness of the importance and diversity of bees. She is the author of *Dancing with Bees*. Twitter: @B_Strawbridge

CHARLOTTE STRAWBRIDGE is a contemporary artist, based in Edinburgh. Inspired by the flora and fauna of the British countryside, she uses oils to create paintings of nature and movement. Instagram: @DearUniverse_cs

DAVID STROUD is obsessed with Greenland White-fronted Geese and has studied them throughout their international distribution for over 30 years, coordinating research and advocating necessary conservation.

JANE TAVENER lives in the Malvern Hills, works in biological recording and is an activist with Extinction Rebellion. She creates art inspired by her adventures in the more-than-human world. Twitter: @jane.tavener

ANNA TERREROS-MARTIN is a children's book author and illustrator and studied an MA in Children's Book Illustration. When away from her desk, Anna loves bird watching and walking in the Peak District. Twitter: @anna_terreros

PAUL THORBY is a structural technician originally on the drawing board and now exclusively CAD. Inspired by his grandfather, he is a keen birder and artist living on the edge of the Cambridgeshire Fens.

MIKE TOMS is a naturalist who writes books, articles and poems about birds and the natural world, including two in the Collins *New Naturalist* series. Twitter: @miketoms

ISABELLA TREE is an award-winning author, married to the conservationist Charlie Burrell. Her best-selling book *Wilding* tells the story of their pioneering rewilding experiment at Knepp Estate in West Sussex. Twitter: @isabella_tree

JON TREMAINE is a Cornish wildlife artist who creates amazing pen and ink composite art that defies definition. His drawings are an amalgam of images; wildlife and flora entwine adding intrigue to this beautiful and unique art form. www.jontremaine.com

ESTHER TYSON RCA is a painter who predominantly works in oil. She currently lives and works in the south Peak District, where she combines a studio and observational outdoor practice. www.esthertyson.co.uk Instagram: @esthertyson

JULIET VICKERY is a conservation scientist whose work in the UK and abroad has focussed on understanding the causes of bird declines and how to reverse them.
Twitter: @juliet_vickery

ROSIE VILLIERS-STUART is a wildlife and landscape painter living in Northumberland. She works mostly in the field and loves painting on bird cliffs, rocky coasts and estuaries. Instagram: @rosievilliersstuart

RUTH WALKER is a graphite artist based in Norfolk. She combines her passions for wildlife and art by working at the BTO by day and drawing by night.
Instagram: @ruthwalkerart www.ruthwalkerart.co.uk

FLORA WALLACE is a botanical ink maker, illustrator and ceramicist. Her practices are intertwined with and inspired by the natural world. Her Nightingale was painted with ink made from Black Walnuts, Oak gall and iron water. www.florawallace.com
Instagram: @florascottwallace

PAUL WARD is an artist who spends most of his time hunched over a desk working. For the rest of the time, he can be found gazing up at trees and picking up insects.
Instagram: @paulwardart

BEN WATT is a musician, writer and birder, best known as one half of Everything But The Girl. In 2021 he founded environmental pressure group Cool Oak. Twitter: @ben_watt

RUTH WEAVER has been fascinated with birds, especially raptors, since childhood. She is now a professional artist, combining her passion for birds to create highly detailed coloured pencil drawings. www.ruths-portraits.com

STEVE WILLIS is a birder and naturalist based in Scotland. He has a background as a Countryside Ranger, walking and wildlife guide, and has worked on various species conservation projects. Twitter: @squirrelsteve2

DAVID WOOD is a reserve manager for the RSPB on the Isle of Islay, birder and occasional artist creating linocuts for his wife's printmaking business, Islay Prints.
www.islayprints.co.uk

TIM WOOTTON is a life-long birder and artist, author of *Drawing & Painting Birds* (Crowood Press), winner of the Swarovski Artist of the Year, and a member of Society of Wildlife Artists.

JO WRIGHT is a wildlife printmaker and keen naturalist. She is inspired by animal encounters and travels in the natural world and all her work has a story to tell.
See more on Instagram: @jowright.printmaker

Index

Arctic Skua 88
Balearic Shearwater 21
Bewick's Swan 24
Black Grouse 22
Black-tailed Godwit 70
Capercaillie 20
Cirl Bunting 152
Common Scoter 32
Corn Bunting 150
Corncrake 52
Cuckoo 50
Curlew 68
Dotterel 60
Dunlin 74
Fieldfare 122
Golden Oriole 157
Goldeneye 34
Grasshopper Warbler 112
Greenfinch 142
Grey Partridge 16
Hawfinch 140
Hen Harrier 92
Herring Gull 84
House Martin 114
House Sparrow 132
Kittiwake 82
Lapwing 64
Leach's Storm Petrel 54
Lesser Redpoll 148
Lesser Spotted Woodpecker 96

Linnet 146
Long-tailed Duck 28
Marsh Tit 102
Marsh Warbler 108
Merlin 98
Mistle Thrush 120
Montagu's Harrier 94
Nightingale 128
Pochard 38
Ptarmigan 18
Puffin 90
Purple Sandpiper 76
Red-backed Shrike 100
Red-necked Grebe 42
Red-necked Phalarope 80
Redpoll 148
Ring Ouzel 124
Ringed Plover 62
Roseate Tern 86
Ruff 72
Savi's Warbler 110
Scaup 40
Shag 58
Skylark 106
Slavonian Grebe 44
Smew 36
Spotted Flycatcher 126
Starling 118
Swift 48
Tree Pipit 136

Tree Sparrow 134
Turtle Dove 46
Twite 144
Velvet Scoter 30
Whimbrel 66
Whinchat 130
White-fronted Goose 26
Willow Tit 104
Wood Warbler 116
Woodcock 78
Yellow Wagtail 138
Yellowhammer 154

Contributors

Laura Andrew 142	Brin Edwards 131	Annie Irving 40
Mark Andrews 109	Rosie Ellis 66	Richard Jarvis 127
Graham Appleton 74	Lissa Evans 120	Kit Jewitt 5
Holly Astle 87	John Fanshawe 58	Emily Joáchim 94
Jeff Baker 111	Dr Annette Fayet 86	Shauna Laurel Jones 28
Dawn Balmer 80	Jake Fiennes 16	Kabir Kaul 38
David Bennett 41	Jonnie Fisk 32	Dr Amir Khan 143
Bo Beolens 108	Beatrice Forshall 47	Mike Langman 33
Tony Blythe 23	Jess French 72	Lucy Lapwing 146
Tessa Boase 64	Federico Gemma 21	Sam Lee 128
Katherine Booth Jones 85	Josie George 102	Olivier Leger 57
Kate Bradbury 132	Jennifer Gill 70	M. G. Leonard 118
Brian Briggs 110	David Gray 68	Nora Ligus 43
Julie Brominicks 148	James Green 115	Steven Lovatt 124
Adam Buxton 106, 107	Malcom Green 50	Richard Mabey 48
Steve Cale 51	Indy Kiemel Greene 144	Benedict Macdonald 20
Hayley Chan 121	Mark Gurney 61	Andrew Mackay 83
Denis Chavigny 137	David Hall 91	Myles Mansfield 149
Faith Chevannes 105	Paul Harfleet 45	Melanie Mascarenhas 69
Miles Cluff 117	Philip Harris 93	Scott Mayson 141
Daniel Cole 59	Celia Hart 25	Megan McCubbin 78
Pete Cooper 126	Andrew Haslen 17	James McCallum 79
Mackenzie Crook 97	Ren Hathway 55	Tom McKinney 42
Nicola Davies 112	John Hatton 37	Harriet Mead 49
Alison Deegan 155	Alicia Hayden 60	Jim Moir 67
Heather Devey 82	Nick Hayes 27	Gary Moore 116
Jamey Douglas 147	Lisa Hooper 81	Charlie Moores 92
Aaron Duff 88	Alex Horne 98	Helen Musselwhite 103
Adrian Dutt 153	Martin Hughes-Games 140	David Noble 114
Dr Mark Eaton 9, 76	Liz Humphreys 84	Jamie Normington 134

173

Kathryn O'Kell 113	Tom Stewart 30
Elizabeth Parker 130	Philip Strange 152
James Parry 150	Brigit Strawbridge 122
Peter Partington 29	Charlotte Strawbridge 123
Ruth Peacey 22	David Stroud 26
Nik Pollard 71	Jane Tavener 101
Karine Polwart 52	Anna Terreros-Martin 73
Tim Pond 35	Paul Thorby 139
Greg Poole 151	Mike Toms 7, 11, 156
Jim Pratt 44	Isabella Tree 46
Emma Price 135	Jon Tremaine 95
Eileen Rees 24	Esther Tyson 133
Sara Rhys 53	Juliet Vickery 138
Laurence Rose 104	Rosie Villiers-Stuart 63
Will Rose 157	Ruth Walker 77
Ben Rothery 119	Flora Wallace 129
Dan Rouse 61	Paul Ward 19
Stephen Rutt 34	Ben Watt 36
Lee Schofield 136	Ruth Weaver 99
José Antonio Sencianes 31	Steve Willis 18
Stewart Sexton 125	David Wood 145
Megan Shersby 154	Tim Wootton 89
Celia Smith 75	Jo Wright 65
Ken Smith 96	
Linda Smith 96	
Manuel Sosa 39	
Beccy Speight 90	
Katie Stacey 100	
Paul Stancliffe 54	
David Steel 56